NELSON
VICmaths

VCE UNITS ① + ②

mathematical methods 11

mastery workbook

Greg Neal
Sue Garner
George Dimitriadis
Toudi Kouris
Stephen Swift

Nelson VICmaths Mathematical Methods 11 Mastery Workbook
1st Edition
Greg Neal
Sue Garner
George Dimitriadis
Toudi Kouris
Stephen Swift
ISBN 9780170464093

Publisher: Dirk Strasser
Additional content created by: ansrsource
Project editor: Alan Stewart
Series cover design: Leigh Ashforth (Watershed Art & Design)
Series text design: Rina Gargano (Alba Design)
Series designer: Nikita Bansal
Production controller: Karen Young
Typeset by: MPS Limited

Any URLs contained in this publication were checked for currency during the production process. Note, however, that the publisher cannot vouch for the ongoing currency of URLs.

Acknowledgements
TI-Nspire: Images used with permission by Texas Instruments, Inc
Casio ClassPad: Shriro Australia Pty. Ltd.

For product information and technology assistance,
in Australia call **1300 790 853**;
in New Zealand call **0800 449 725**

For permission to use material from this text or product, please email
aust.permissions@cengage.com

ISBN 978 0 17 046409 3

Cengage Learning Australia
Level 7, 80 Dorcas Street
South Melbourne, Victoria Australia 3205

Cengage Learning New Zealand
Unit 4B Rosedale Office Park
331 Rosedale Road, Albany, North Shore 0632, NZ

For learning solutions, visit **cengage.com.au**

Printed in China by 1010 Printing International Limited.
1 2 3 4 5 6 7 26 25 24 23 22

9780170464093

Contents

Probability and counting   **103**

The circular functions   173

9

Differentiation   190

10

Trigonometric and exponential equations

204

Applications of differentiation

221

To the student

Nelson VICmaths is your best friend when it comes to studying Mathematical Methods in Year 11. It has been written to help you maximise your learning and success this year. Every explanation, every exam hack and every worked example has been written with the exams in mind.

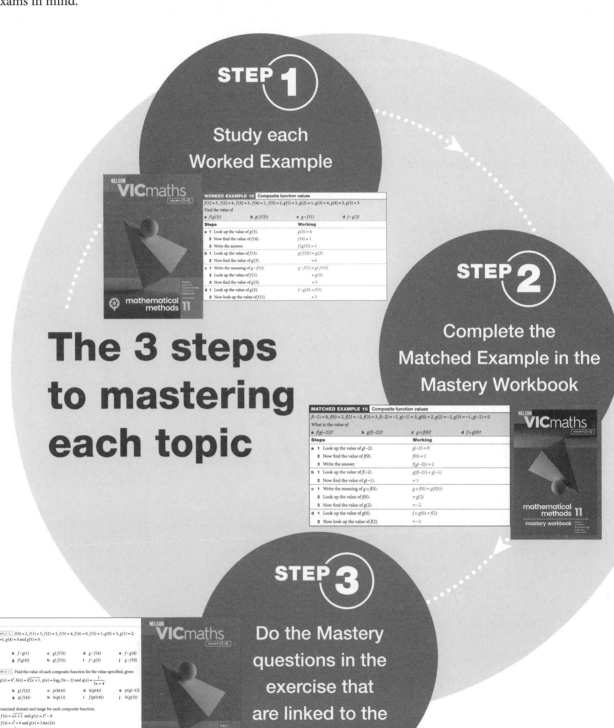

The 3 steps to mastering each topic

STEP 1 Study each Worked Example

STEP 2 Complete the Matched Example in the Mastery Workbook

STEP 3 Do the Mastery questions in the exercise that are linked to the Worked Example

EQUATIONS AND INEQUALITIES

MATCHED EXAMPLE 1	Index laws 1

Simplify the expression $a^4 \times b^3 \div b^6 \times a^{-6} \times b^4$.

Steps	Working
1 Write the expression and group the terms with the same base.	
2 Use index laws to simplify.	
3 Simplify indices.	

p. 4

MATCHED EXAMPLE 2 | Index laws 2

Simplify the expression $81^{\frac{3}{4}} \div 27^{\frac{1}{3}} \times 7^{\frac{1}{5}} \times 7^{\frac{4}{5}}$.

Steps	Working
1 Write the expression and simplify the bases as prime numbers if possible.	
2 Use index laws to simplify.	

MATCHED EXAMPLE 3 | Exponential equations

Solve the equation $6^{-x} \div 6^{2-4x} = 36^x$ for x.

SB

p. 5

Steps	Working
1 Express each term with the base 6 and simplify using index laws.	
2 Equate powers.	
3 Solve for x.	

MATCHED EXAMPLE 4	Binomial expansions

Expand and simplify the expression $(x + 4)(3x - 2)$.

Steps	Working

1 Use the distributive law to $(x + 4)(3x - 2)$.

Or use the **FOIL method.**

2 First: $x \times 3x = 3x^2$

Outer: $x \times -2 = -2x$

Inner: $4 \times 3x = 12x$

Last: $4 \times -2 = -8$

MATCHED EXAMPLE 5 | Perfect squares

Expand the expression $(2x + 1)^2$.

p. 7

Steps	Working
1 Use the distributive law or the FOIL method to expand $(2x + 1)^2$.	
2 Or use the formula $(a + b)^2 = a^2 + 2ab + b^2$.	

MATCHED EXAMPLE 6 Difference of two squares

Expand the expression $(3x + 6)(3x - 6)$.

Steps	Working
1 Use the FOIL method to expand $(3x + 6)(3x - 6)$.	
2 Alternatively, use the difference of two squares formula $(a + b)(a - b) = a^2 - b^2$.	

MATCHED EXAMPLE 7 | Factorising perfect squares

Factorise the perfect square $9x^2 - 12x + 4$.

SB

p. 9

Step	Working
Use the formula $(a - b)^2 = a^2 - 2ab + b^2$.	

MATCHED EXAMPLE 8 | Factorising the difference of squares

Factorise the expression $64x^2 - 9$.

Steps	Working
Use the difference of two squares formula $(a + b)(a - b) = a^2 - b^2$ in the expression $64x^2 - 9$.	

MATCHED EXAMPLE 9	Factorising by grouping in pairs

p. 10

Factorise each expression.

a $x^3 - 2x^2 - 9x + 18$

b $3x^2 + 9x - x - 3$

Steps	Working
a 1 Factorise the first 2 terms, then the last 2 terms.	
2 Factorise again using the common factor.	
3 Factorise $x^2 - 9$.	
b 1 Factorise the first 2 terms, then the last 2 terms.	
2 Factorise again using the common factor.	

MATCHED EXAMPLE 10 | Solving linear equations

Solve $4(x - 2) = 7x - 1$ for x.

Steps	Working
1 Expand the brackets.	
2 Add 1 on both sides.	
3 Subtract $4x$ from both sides.	
4 Divide both sides by 3.	

9780170464093

MATCHED EXAMPLE 11 | Roots of a graph

Find the roots of the graph of $y = 3x^2 - 15x$.

p. 12

Steps	Working
1 Factorise the expression $3x^2 - 15x$.	
2 Solve the equation $3x^2 - 15x = 0$.	
3 Express the roots of the graph in coordinate form.	
4 Sketch the graph of $y = 3x^2 - 15x$ labelling the roots.	

MATCHED EXAMPLE 12 | Factorising a monic quadratic expression

Factorise the expression $x^2 - 5x + 6$.

Hence, solve the quadratic equation $x^2 - 5x + 6 = 0$.

Steps	Working
1 Write 2 pairs of brackets beginning with 'x'.	
2 Find 2 numbers with a sum of -5 and a product of $+6$. Look for factors of $+6$, positive and negative.	
3 Solve the equation.	

MATCHED EXAMPLE 13 | Factorising a non-monic quadratic expression

Factorise the expression $3x^2 + 5x - 2$.

Hence, solve the quadratic equation $3x^2 + 5x - 2 = 0$.

Steps	Working
1 Find 2 numbers with a sum of +5 and a product of $3 \times (-2) = -6$ Use these numbers to split the middle term ($+ x$).	
2 Factorise in pairs.	
3 Take out the common factor.	
4 Solve the equation.	

p. 16

MATCHED EXAMPLE 14 | Completing the square

Solve the equation $2x^2 + 20x = -19$ by completing the square.

Steps	Working
1 Write the equation in general form: $ax^2 + bx + c = 0$.	
2 Take out the common factor of a.	
3 Halve the coefficient of x and then square this number.	
4 Add and then subtract the number found in the previous step.	
5 Recognise the perfect square.	
6 Divide both sides of the equation by 2.	
7 Alternatively, use difference of squares.	

MATCHED EXAMPLE 15 | Using the quadratic formula

Solve the equation $3x^2 = 4x + 5$ using the quadratic formula.

Steps	Working
1 Write in general form: $ax^2 + bx + c = 0$.	
2 Identify a, b, c.	
3 Substitute into the quadratic formula $$x = \frac{-b \pm \sqrt{b^2 - 4ac}}{2a}$$ simplifying where necessary.	

MATCHED EXAMPLE 16 | The discriminant

a Show that the quadratic equation $3x^2 + 6x + 4 = 0$ has no real roots.

b Find the value(s) of k for which the equation $3x^2 + 4x + k = 0$ has real roots.

Steps	Working
a 1 Write the discriminant, $\Delta = b^2 - 4ac$, and substitute $a = 3$, $b = 6$, $c = 4$.	
2 Show that the quadratic equation $3x^2 + 6x + 4 = 0$ has no real roots.	
b 1 Write the discriminant condition.	
2 Make $\Delta \geq 0$ for real roots and solve for k.	

MATCHED EXAMPLE 17 | Literal equations

Make h the subject of the formula $A = 2\pi r^2 + 2\pi rh$.

Steps	Working
Solve the equation for h. Rearrange using inverse operations.	

SB

Using CAS 3:
Transposing
formula
p. 21

MATCHED EXAMPLE 18 Simultaneous equations using graphs

Use graphs to solve the simultaneous equations $4x - 6y - 6 = 0$ and $3y = 9 - 2x$.

Steps	Working
1 Graph $4x - 6y - 6 = 0$ and $3y = 9 - 2x$.	
2 Read the solution point from the graph.	

9780170464093

Using an algebraic method, solve the simultaneous equations $2x - 7 = y$ and $9 = 3x - 2y$.

SB
p. 23

Steps	Working
1 Elimination method	
Rearrange the equations to get matching terms and add or subtract to eliminate x or y.	
2 Substitute $x = 5$ into the easiest equation, say [eqn 1], to find y.	
3 Substitution method	
Substitute $y = 2x - 7$ into $9 = 3x - 2y$ and solve to find x.	
4 Substitute $x = 5$ into $2x - 7 = y$ to find y.	

SB

Using CAS 4:
Simultaneous
equations
p. 24

Using CAS 5:
Solutions up to
four unknowns
p. 25

MATCHED EXAMPLE 20 | Finding a constant in simultaneous equations

Find the value(s) of k such that the simultaneous equations $(k-2)x + 3y = 6$ and $kx - 6y = k - 1$ have no solution, where k is a real constant.

Steps	Working
1 Find the gradient of both linear equations by rearranging the equations in the form $y = mx + c$.	
2 If the equations have no solution, then the lines must be parallel, so their gradients must be equal but their y-intercepts must be different (so that they are not the same line).	
3 Substitute $k = \dfrac{4}{3}$ to check that they are NOT the same line.	
4 Write the answer.	

MATCHED EXAMPLE 21 | Inequalities

Solve for x in the inequality $\dfrac{3x+1}{4} \le 2x - 3$.

p. 28

Steps	**Working**
Solve using inverse operations, cross-multiplying first.	

Using CAS 6:
Solving
inequalities
p. 29

CHAPTER

2 FUNCTIONS AND GRAPHS

MATCHED EXAMPLE 1	Mappings and ordered pairs

The mapping below shows the relation M.

a Write M as a set of ordered pairs and state the domain and range.

b Show M as a table.

c State the domain and range of M.

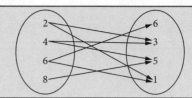

Steps	Working
a Write the ordered pairs.	
b Write the ordered pairs in a table.	
c 1 The domain is the set of first numbers.	
2 The range is the set of second numbers.	

MATCHED EXAMPLE 2 | Relations and functions

State whether each relation is a function.

a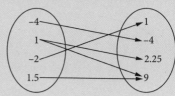

b $y = 2x + 1$ **c** $y^2 = x^2 + 16$ **d**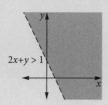

Steps	Working
a 1 Write the ordered pairs.	
2 Write the conclusion.	
b 1 Do any ordered pairs have the same x-value?	
2 Write the conclusion.	
c 1 Do any ordered pairs have the same x-value?	
2 Write the conclusion.	
d 1 Do any ordered pairs have the same x-value?	
2 Write the conclusion.	

MATCHED EXAMPLE 3 Relations and function graphs

State whether each of the following shows the graph of a function.

a

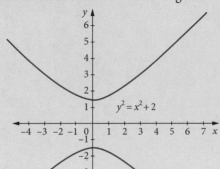

$y^2 = x^2 + 2$

b

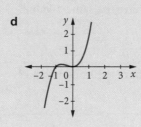

$y = \sqrt{\pm x - 1}$

c

d

Steps	Working
a 1 Do any points have the same x-value?	
2 Write the conclusion.	
b 1 Do any points have the same x-value?	
2 Write the conclusion.	
c 1 Do any points have the same x-value?	
2 Write the conclusion.	
d 1 Do any points have the same x-value?	
2 Write the conclusion.	

MATCHED EXAMPLE 4 | Straight lines

A botanist wanted to find the initial growth rate of a plant from seed. She soaked a seed in water overnight and placed it in a ziplock bag to observe the root length every two days.

Draw a graph and determine the relationship between the time and root length of the seed.

Time (days)	0	2	4	6	8	10
Root length (cm)	0	0.8	2	2.9	3.7	4.7

Steps	Working

1 Plot the points on suitable graph paper.

The points appear to be in a line, so draw a line of best fit.

2 Choose 2 easy points on the line.

3 Calculate the gradient of the line.

4 Write the equation using $y = mx + c$ and give the answer.

MATCHED EXAMPLE 5 | Plotting a linear function

The total cost for running a pizza night at a sports club by a group of students is $220.

Two hundred pizza slices are to be sold for $4 each slice.

a Write the function for the profit made when n slices are sold.

b How many slices must be sold to make a profit?

c Draw a graph of the profit function.

d How much profit is made if all 200 slices are sold?

Steps	Working
a Profit = sales − cost	
b 1 Find $P(n) > 0$.	
2 Write the answer.	
c 1 Choose some values of n and calculate $P(n)$.	
2 Draw the graph.	
d 1 Find $P(200)$.	
2 Write the answer.	

MATCHED EXAMPLE 6 | Graphing a quadratic function

Graph $f(x) = 4x - x^2 - 5$ from $x = -2$ to $x = 6$

Steps	Working
1 Make a table of values.	
2 Plot the points and join them with a smooth curve.	

SB

Using CAS 1:
Sketching graphs
p. 45

MATCHED EXAMPLE 7 Features of a graph with no specific rule

The graph of a function is shown below. Identify and approximately locate its significant features.

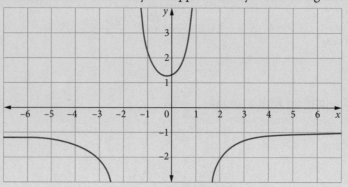

Steps	Working
1 Where does the graph cross the y-axis?	
2 Where does it cross the x-axis?	
3 Are there any maxima or minima?	
4 Are there any asymptotes?	
5 Are there any points of inflection?	

MATCHED EXAMPLE 8 | Existence of an inverse function

State whether each function has an inverse function. If not, suggest how the domain could be restricted so that it does.

p. 47

a $(1, 3), (3, 6), (2, 4), (6, 18), (8, 21)$

b $y = x^2 - 1$

c $f(x) = x + 3$

Steps	Working
a 1 Check the y-values for duplicates.	
2 State the conclusion.	
b 1 Check the y-values for duplicates.	
2 Find the turning point.	
3 Suggest a restriction.	
c 1 Check the y-values for duplicates.	
2 State the conclusion.	

MATCHED EXAMPLE 9 | Function and inverse function graphs

Find the inverse function of each of the following functions and sketch both, together with the line $y = x$.

a $f: (2, 1), (-2, 3), (3, 2), (-3, 4), (1, 0)$

b $f(x) = 3x + 12$

Steps	Working
a **1** Write the inverse function.	
2 Plot the graphs in different colours. Draw $y = x$.	
b **1** Swap x and y.	
2 Make the new y the subject.	
3 Write the inverse function.	
4 Find some points on the graphs.	
5 Draw the graphs in different colours.	

MATCHED EXAMPLE 10	Maximal domain of a function

What is the maximal domain of each of these functions?

a $y = \sqrt{x^2 - 1}$ **b** $h(x) = \dfrac{1}{2x - 6}$

Steps	Working
a 1 When is $\sqrt{x^2 - 1}$ defined?	
2 Write the answer.	
b 1 When is $\dfrac{1}{2x - 6}$ defined?	
2 Write the answer.	

2

MATCHED EXAMPLE 11 Range of a linear function with a restricted domain

What is the range of $g: [-3, 9) \to R, g(x) = 2x - 3$?

Steps	Working
1 $g(x)$ is a line between $x = -3$ and $x = 9$, so find the y values at the endpoints.	
2 Write the range.	

MATCHED EXAMPLE 12	Range of a quadratic function	

Find the range of the function given by $f: R \rightarrow R, f(x) = x^2 - 4x + 8$.

SB

p. 54

Steps	Working
1 Complete the square.	
2 State the turning point.	
3 Sketch the graph.	
4 Write the range.	

MATCHED EXAMPLE 13 | Range of a restricted function

Find the range of the function given by $f: (1, 15] \to R, f(x) = \dfrac{15-x}{x-1}$.

Steps	Working
1 Check values close to 1.	
2 Check values close to and equal to 15.	
3 Check other values.	
4 Sketch the graph.	
5 Write the range.	

MATCHED EXAMPLE 14 | Domain and range

What is the domain D of $h(x) = 9 + 3x$ if the range is $[9, 60)$?

Steps	Working
1 Solve for one endpoint.	
2 Solve for the other endpoint.	
3 Write the answer.	

SB

p. 59

MATCHED EXAMPLE 15 | Equation of a straight line

Find the equation of a straight line that

a passes through (1, 4) and has gradient 3.

b has y-intercept 3 and gradient -1.

c passes through (1, -1) and (2, 3).

d has x-intercept 3 and y-intercept 4.

Steps	Working
a 1 Write the point-gradient form. **2** Substitute values. **3** Simplify and express in general form.	
b 1 Write the point-gradient form. **2** Substitute $m = -1$ and the point (0, 3). **3** Simplify and express in general form.	
c 1 Find the gradient. **2** Write the point-gradient form. **3** Substitute the 'simpler' point, simplify and express in general form.	
d 1 Find the gradient. **2** Write the point-gradient form. **3** Substitute the 'simpler' point, simplify and express in general form.	

9780170464093

MATCHED EXAMPLE 16 | Using perpendicular lines

The linear function $x + 2y = 6$ has x and y intercepts A and B. C is the point that divides AB in the ratio $1 : 2$. Show that the perpendicular line to AB through C also passes through the point $O(5, 3)$.

Steps	Working
1 Sketch a diagram.	
2 Find A and B.	
3 Find C.	
4 Find the gradient of AB.	
5 Find the gradient of the perpendicular line, CD.	
6 Find the equation of CD.	
7 Substitute O in the LHS.	
8 State the conclusion.	

MATCHED EXAMPLE 17 | Completing the square

Find the turning point of $f(x) = 6x - 2x^2 - 13$.

Steps	Working
1 Separate the constant and take the coefficient of x^2 as a factor.	
2 Complete the square.	
3 Simplify the constant terms.	
4 State the turning point.	

MATCHED EXAMPLE 18	Graphing $y = a(x - h)^2 + k$

Sketch the graph of $f(x) = -3(x - 1)^2 - 3$.

Steps	**Working**
1 Compare to $y = x^2$.	
2 State the shape.	
3 State the y-intercept.	
4 Sketch the graph, giving the coordinates of all important points.	

2

MATCHED EXAMPLE 19 | Sketching a factorable quadratic function

Sketch the graph of $f(x) = 3 - x^2 - 2x$.

Steps	Working
1 Factorise if possible and state the zeros.	
2 The zeros are symmetrical about the turning point.	
3 State the turning point.	
4 State the shape.	
5 State the y-intercept.	
6 Sketch the graph.	

MATCHED EXAMPLE 20 | Sketching a non-factorable quadratic function

Sketch the graph of $y = 2x - 5 + x^2$.

Steps	Working
1 The quadratic won't factorise, so complete the square.	
2 Compare to $y = x^2$.	
3 State the shape.	
4 Find the zeros. The approximate value will help you sketch the function, but put the exact values on the sketch.	
5 Sate the y-intercept.	
6 Sketch the graph.	

MATCHED EXAMPLE 21 | Values of a quadratic coefficient

Find the values of b such that $f(x) = x^2 + bx - 2$ and $y = 5x - b$ have two intersections.

Steps	Working
1 Write the equation for intersections.	
2 Find the discriminant using $\Delta = b^2 - 4ac$.	
3 Use the condition for two intersections.	
4 Find other criteria of the graph of Δ.	
5 Sketch the graph of Δ.	
6 State the values of b such that $\Delta > 0$.	
7 State the answer.	

9780170464093

POLYNOMIALS

MATCHED EXAMPLE 1	Polynomial expressions and their features

Determine whether each expression is a polynomial, giving reasons if it is not. State the degree, leading term and coefficients of any polynomial found.

a $x + \sqrt{x} - 10 + x^2$ **b** $2x^3 - 8 + 3x^2$

Steps	Working
a Write the expression with descending powers. Not all of the powers are positive whole numbers, so it is not a polynomial.	
b 1 Write the expression with descending powers of x. All of the powers are positive whole numbers, so it is a polynomial.	
2 Identify the degree, leading term and coefficients.	

SB

p. 80

MATCHED EXAMPLE 2 Simplifying polynomials

If $P(x) = 3x^4 + 2x^2 - x - 8$ and $Q(x) = 3x^3 - 4x + 10$, find $P(x) + Q(x)$ and state its degree.

Steps	Working
1 Substitute for $P(x)$ and $Q(x)$.	
2 Add or subtract like terms to simplify.	

MATCHED EXAMPLE 3 Substituting values into polynomials

a Given that $P(x) = 2x^2 + 10x + 4$, find:

  i $P(-1)$ **ii** $P(a + 2)$

b x when $P(x) = -4$

Steps	Working
a **i** Substitute $x = -1$ into $P(x)$ and evaluate.	
ii Substitute $x = a + 2$ into $P(x)$ and simplify.	
b Substitute $P(x) = -4$ and solve for x.	

SB
Using CAS 1:
Defining and
evaluating
polynomials p. 81

3

MATCHED EXAMPLE 4 | Finding unknown variables

If $P(x) = 2x^3 + 5x^2 - mx - 13$ and $P(2) = 32$, then find the value of m.

Steps	Working
1 Substitute $(2, 32)$ into $P(x)$.	
2 Solve for m.	

MATCHED EXAMPLE 5 | Modelling with polynomials

The height, h metres, of a cannonball fired from a hill above the ground after t seconds is modelled by the equation $h(t) = -1.8t^2 + 15t + 250$.

a Find its height above the ground after 5 seconds.

b When will the ball hit the ground? Give your answer correct to two decimal places.

Steps	Working
a Substitute $t = 5$ into $h(t)$.	
b 1 This will happen when the height is 0. Solve the equation $h(t) = 0$.	
2 Check the feasibility of the solutions.	

3

p. 84

MATCHED EXAMPLE 6 | Equating coefficients

If $4x^2 - 8x + 7 = a(x - 1)^2 + b$ for all x, find the values of a and b.

Steps	Working
1 Expand the RHS.	
2 Let LHS = RHS.	
3 Equate coefficients.	

Divide $P(x) = 2x^3 - 3x^2 - x + 6$ by $d(x) = x - 1$ and hence express $P(x)$ in the form $d(x) \times Q(x) + R$.

Steps	**Working**
1 Divide the highest power of x from $(x - 1)$ into the highest power of $(2x^3 - 3x^2 - x + 6)$. This is $2x^3 \div x$, which gives $2x^2$. Put this up the top in the x^2 place. Work out how much of $(2x^3 - 3x^2)$ has been accounted for by doing $(x - 1) \times 2x^2 = 2x^3 - 2x^2$. Put this under the $(2x^3 - 3x^2)$ and subtract to work out how many x^2 is left $(-x^2)$. Then, bring down the $- x$.	
2 x into $-x^2$ goes $-x$ times. Put the $-x$ in the x place, above the $-x$. $(x - 1) \times (-x) = -x^2 + x$. Put this underneath the $-x^2 - x$ and subtract to get $-2x$. Bring down the 6.	
3 x into $-2x$ goes -2 times. Put the -2 above 6. $(x - 1) \times -2 = -2x + 2$. Put this underneath the $-2x + 6$ and subtract to get 4.	
4 Write the answer in the form $P(x) = d(x) \times Q(x) + R$.	

MATCHED EXAMPLE 8	Division of a polynomial by a non-linear divisor

Divide $P(x) = 3x^3 - 4x^2 + 2x - 1$ by $d(x) = 3x^2 + 2$ to find the quotient and remainder.

Steps	**Working**
1 Divide the highest power of x from $(3x^2 + 2)$ into the highest power of $(3x^3 - 4x^2 + 2x - 1)$. This is $3x^3 \div 3x^2$, which gives x. Put this up the top in the x place. Work out how much of $(3x^3 - 4x^2)$ has been accounted for by doing $(3x^2 + 2) \times x = 3x^3 + 2x$. Put this under the $(3x^3 - 4x^2)$ and subtract to work out how many x^2 is left. Then, bring down the -1.	
2 $3x^2$ into $-4x^2$ goes $-\dfrac{4}{3}$ times. Put the $-\dfrac{4}{3}$ in the units place, above the -1. $(3x^2 + 2) \times \left(-\dfrac{4}{3}\right) = -4x^2 - \dfrac{8}{3}$ Put this underneath the $-4x^2 - 1$ and subtract to get $\dfrac{5}{3}$.	
3 Write the answer.	

MATCHED EXAMPLE 9 Division of a polynomial using synthetic division

Using synthetic division, divide $P(x) = x^3 - 2x^2 + 5x - 1$ by $d(x) = 4x - 1$, and hence express $P(x)$ in the form $d(x) \times Q(x) + R$.

Steps	Working
1 Set $d(x) = 0$ and solve for x.	
2 Place the solution to $d(x) = 0$, $\dfrac{1}{4}$, in the top left-hand corner of the first row, to the left of the separation line. Place the coefficients of the terms in $P(x)$ in the top row to the right of the separation line, in order of their decreasing power.	
3 Bring the first coefficient, 1, down. $\dfrac{1}{4} \times 1 = \dfrac{1}{4}$ and place this underneath -2, and add to get $-\dfrac{7}{4}$.	
4 The rest of the values in the second row are found by multiplying each value in the bottom row by the $\dfrac{1}{4}$ from $(x - \dfrac{1}{4})$. The values in the row are column totals, which make up the coefficients of the quotient, the last value being the remainder.	
5 Express answer in the form $P(x) = d(x) \times Q(x) + R$. In order to do this, $\left(x - \dfrac{1}{4} \right)$ must be factorised by taking out a factor of $\dfrac{1}{4}$ to get $\dfrac{1}{4}(4x - 1)$. This factor of $\dfrac{1}{4}$ is then multiplied to the second bracket.	
6 Write the answer in the form $P(x) = d(x) \times Q(x) + R$.	

MATCHED EXAMPLE 10 | Division by inspection

Given $P(x) = 3x + 4$ and $d(x) = x - 4$, use division by inspection to express $P(x)$ in the form $Q(x) + \dfrac{R}{d(x)}$.

Steps	Working
1 Write the numerator $3x + 4$ as $3(x - 4) + 4 + 12$, which simplifies to $3(x - 4) + 16$.	
2 Split the expression into two partial fractions.	
3 Simplify.	

Find the remainder when $3x^3 + 2x^2 + 2x - 3$ is divided by $3x - 1$.

SB

p. 92

Steps	Working
1 Write down $P(x)$.	
2 The remainder when $P(x)$ is divided by $3x - 1$ is $P\left(\dfrac{1}{3}\right)$. Evaluate $P\left(\dfrac{1}{3}\right)$.	
3 Write the answer.	

3

MATCHED EXAMPLE 12 | Using the remainder theorem to find a coefficient

Evaluate m if the remainder is 6 when dividing $3x^4 + mx + 6$ by $x + 2$.

Steps	Working
1 Write down $P(x)$.	
2 The remainder when $P(x)$ is divided by $x + 2$ is $P(-2)$.	
3 The remainder is 6, so $P(-2) = 6$.	
4 Solve the equation for m.	

MATCHED EXAMPLE 13 Using the remainder theorem to find coefficients

$f(x) = x^3 + ax^2 + bx - 4$, where a, b are constants.

a Given that when $f(x)$ is divided by $(x + 1)$, the remainder is -9. Show that $a - b = -4$.

b Given also that when $f(x)$ is divided by $(x - 2)$, the remainder is -6. Find the value of a and b and hence determine the polynomial $f(x)$.

Steps	Working
a 1 Write down $f(x)$.	
2 The remainder when $P(x)$ is divided by $x + 1$ is $P(-1)$.	
3 The remainder is -9, so $P(-1) = -9$.	
4 Simplify the equation and show the result given.	
b 1 The remainder when $P(x)$ is divided by $x - 2$ is $P(2)$.	
2 The remainder is -6, so $P(2) = -6$.	
3 Set up a pair of simultaneous equations.	
4 Solve for a and b.	
5 Write the polynomial $f(x)$ using a and b.	

SB

Using CAS 3:
Finding the
remainder of
polynomials
p. 93

MATCHED EXAMPLE 14 | Factorising polynomials using long division 1

a Show that $x - 2$ is a factor of $P(x) = x^3 + 5x^2 - 2x - 24$.

b Using long division, divide $P(x)$ by $x - 2$.

c Factorise $P(x)$ completely.

Steps	Working
a 1 Write down $P(x)$.	
2 To show that $(x - 2)$ to be a factor of $P(x)$, we must prove $P(2) = 0$.	
3 State your answer.	
b 1 Use the process of polynomial long division to divide $x - 2$ into $P(x)$.	
2 State division in the form $P(x) = (x - a)Q(x)$.	
c 1 Factorise $x^2 + 7x + 12$.	
2 Write the answer.	

9780170464093

Factorise $f(x) = x^3 - 3x^2 - 13x + 15$.

Steps	Working
1 Write the polynomial.	
2 Try factors of 15 (i.e., ±1, ±3, ±5, ±15) until the remainder of 0 is found.	
3 Use the process of polynomial long division to divide $x - 1$ into $f(x)$.	
4 Factorise $x^2 - 2x - 15$.	
5 State your answer.	

MATCHED EXAMPLE 16 Factorising polynomials using synthetic division

Factorise $f(x) = x^4 + 4x^3 - x^2 - 16x - 12$.

Steps	Working
1 Write the polynomial.	
2 Try factors of 12 (i.e., $\pm1, \pm2, \pm3, \pm4, \pm6, \pm12$) until the remainder of 0 is found.	
3 Place the -1 in the top left-hand corner of the first row, to the left of the separation line. Bring the first coefficient, 1, down. Place the coefficients of the terms in $P(x)$ in the top row to the right of the separation line, in order of their decreasing power. The values in the second row are found by multiplying each value in the bottom row by the -1 from $(x + 1)$. The values in the bottom row are column totals, which make up the coefficients of the quotient, the last value being the remainder.	
4 Write the answer in the form $P(x) = d(x) \times Q(x) + R$.	
5 Write down the quotient.	
6 To factorise the quotient, repeat the process by first finding a linear factor of $Q(x)$. Try factors of 12 (i.e., $\pm1, \pm2, \pm3, \pm4, \pm6, \pm12$) until the remainder of 0 is found. The rest of the values in the second row are found by multiplying each value in the bottom row by the 2 from $(x - 2)$. The values in the bottom row are column totals, which make up the coefficients of the quotient, the last value being the remainder.	
7 Write down the factorised form of $P(x)$.	
8 Continue factorising by factorising the quadratic polynomial.	

MATCHED EXAMPLE 17 | Factorising polynomials by equating coefficients

Using the method of equating coefficients, factorise $f(x) = 2x^3 - x^2 - 5x - 2$.

Steps	Working
1 Write the polynomial.	
2 Try factors of 2 (i.e., ± 1, ± 2).	
3 Write $f(x)$ in the form of $(x - a)Q(x)$, where $Q(x)$ is a quadratic written in general form and a, b and c are constant coefficients.	
4 Find a by equating coefficients of x^3.	
5 Find c by equating the constant terms.	
6 Find b by equating coefficients of x^2. Note: b can also be found by equating coefficients of x.	
7 State your answer in the form $P(x) = (x - a)Q(x)$.	
8 Factorise completely.	

SB

Using CAS 4:
Finding factors of
polynomials
p. 97

p. 97

MATCHED EXAMPLE 18 | Factorising using the sum and differences of two cubes

Factorise each expression.

a $27x^3 - y^3$

b $24 + 3(x+1)^3$

Steps	Working
a 1 Recognise the difference of two cubes.	
2 Use the formula $a^3 - b^3 = (a - b)(a^2 + ab + b^2)$, where $a = 3x$ and $b = y$.	
b 1 Take out the highest common factor of 3.	
2 Recognise the sum of two cubes.	
3 Use the formula $a^3 + b^3 = (a + b)(a^2 - ab + b^2)$, where $a = 2$ and $b = x + 1$.	
4 Simplify.	

Find the possible rational roots of $P(x) = 2x^3 - 5x^2 + 4x - 8 = 0$.

SB

p. 100

Steps	Working
1 Find the factors of the constant term 8.	
2 Find the factors of the leading coefficient 2.	
3 Write the factors of the constant term over the factors of the leading coefficient.	
4 Write all possible rational roots separately (plus or minus).	

3

SB

p. 101

MATCHED EXAMPLE 20	Factorising using the rational root theorem

$f(x) = 2x^3 - 17x^2 + 5x + 7$ has rational zeros. Factorise $f(x)$.

Steps	Working
1 Find the factors of the constant term 7.	
2 Find the factors of the leading coefficient 2.	
3 Write the factors of the constant term over the factors of the leading coefficient.	
4 Write all possible rational zeros separately.	
5 Instead of checking each possibility, we can narrow the choices by graphing $f(x)$ and using trace to identify possible zeros (x-intercepts).	
6 Use the factor theorem to test.	
7 Factorise $f(x)$ further.	
8 Factor the fraction.	
9 Factorise $(x^2 - 9x + 7)$ by completing the square.	
10 Write the answer.	

MATCHED EXAMPLE 21	Solving equations involving perfect cubes

Solve $3(x - 4)^3 - 192 = 0$.

Steps	**Working**
1 Add 192 to both sides.	
2 Divide both sides by 3.	
3 Take the cube root of both sides.	
4 Add 4 to both sides.	

SB

Using CAS 5:
Solving polynomial
equations
p. 103

p. 103

MATCHED EXAMPLE 22 | Solving quartic equations

Solve $3x^4 - 10x^3 + 9x^2 - 2x = 0$.

Steps	Working
1 Take x out as a common factor.	
2 Use the cubic polynomial and try factors of 2 (i.e., ± 1, ± 2).	
3 **Method 1: Using long division** Divide $P(x)$ by $x - 1$ by using long division.	
Method 2: Using synthetic division Divide $P(x)$ by $x - 1$ by using synthetic division.	
Method 3: By equating coefficients Divide $P(x)$ by $x - 1$ by equating coefficients. Write $f(x)$ in the form of $(x - a)Q(x)$, where $Q(x)$ is a quadratic written in general form $ax^2 + bx + c$ and a, b and c are constant coefficients.	
4 State the polynomial as a product of the factors found so far.	
5 Factorise the quadratic.	
6 Use the null factor law.	
7 Solve the equation.	

MATCHED EXAMPLE 23 | The bisection method

Use the bisection method to find an approximate root of $2x^3 + 5x - 11 = 0$ in the interval $[1, 2]$. Answer correct to two decimal places.

Steps	Working
1 Let $f(x) = 2x^3 + 5x - 11$ and evaluate $f(1)$ and $f(2)$ to check that they are opposite in sign.	
2 Halve the interval and test the function at the halfway point.	
3 Halve the interval again and test the function at the halfway point.	
4 Repeat. Notice that we are getting closer to the root because $f(x)$ is getting closer to 0.	

*Round approximations to two decimal places.

5 Of all the guesses, $f(1.31)$ is closest to 0 (correct to two decimal places).

MATCHED EXAMPLE 24 | Graphing $y = (x - h)^3 + k$

Sketch the graph of $y = (x - 3)^3 + 8$.

Steps	Working
1 Compare $y = (x - 3)^3 + 8$ with $y = x^3$.	
2 Find its x-intercept(s).	
3 Find its y-intercept.	
4 Sketch the graph. It will have the same shape as $y = x^3$.	

MATCHED EXAMPLE 25 | Graphing $y = ax^3 + bx^2 + cx + d$

Sketch the graph of $y = x^3 + x^2 - 12x$, showing all intercepts.

Steps	Working
1 Factorise $y = x^3 + x^2 - 12x$.	
2 Find the x-intercepts.	
3 Find the y-intercept.	
4 Determine the direction of the graph.	
5 Sketch the graph.	

p. 115

MATCHED EXAMPLE 26 | Graphing $y = ax^4 + bx^3 + cx^2 + dx + e$

Sketch the graph of $y = -x^4 + 16x^2$, showing all intercepts.

Steps	Working
1 Factorise $y = -x^4 + 16x^2$.	
2 Find the x-intercepts.	
3 Find the y-intercept.	
4 Determine the direction of the graph.	
5 Sketch the graph.	

SB

p. 115

The function for the graph shown is of the form $y = a(x-b)^4 + c$. Find the equation of the function.

Steps	Working
a 1 Using the turning point, find the value of b and c.	
2 Substitute the y-intercept $(0, -7)$ into the equation.	
3 Write the equation of the function.	

POWER AND INVERSE FUNCTIONS

4

MATCHED EXAMPLE 1	Sketching $y = ax^3$

Sketch the functions

a $y = 0.25x^3$ **b** $f(x) = -2x^3$

Steps	Working
a **1** Write the comparison with $y = x^3$.	
2 State the central point.	
3 State another reference point.	
4 Sketch the graph.	
b **1** Write the comparison with $y = x^3$.	
2 State the central point.	
3 State another reference point.	
4 Sketch the graph.	

MATCHED EXAMPLE 2 | Sketching $y = (x + b)^3$

Sketch the graph of $y = (x - 4)^3$.

SB

p. 135

Steps	Working
1 Write the comparison with $y = x^3$.	
2 State the central point.	
3 State another reference point.	
4 Sketch the graph.	
Find any zeros if possible.	

MATCHED EXAMPLE 3 | Sketching $y = x^3 + c$

Sketch the graph of $f(x) = x^3 - 8$.

Steps	Working
1 Write the comparison with $y = x^3$.	
2 State the central point.	
3 State another reference point.	
4 Sketch the graph.	

Sketch the graph of $f(x) = -x^4 + 5$.

Steps	Working
1 Write the comparison with $y = x^4$.	
2 State the turning point.	
3 Find the zeros if possible.	
4 Sketch the graph.	

MATCHED EXAMPLE 5 | Sketching $y = ax^4$

Sketch the graph of $y = 2x^4$.

Steps	**Working**
1 Write the comparison with $y = x^4$.	
2 State the turning point.	
3 State another reference point.	
4 Sketch the graph.	

MATCHED EXAMPLE 6 Sketching $y = (x + b)^4$

SB
p. 137

Sketch the graph of $f(x) = (x + 3)^4.$

Steps	Working
1 Write the comparison with $y = x^4$.	
2 State the turning point.	
3 Find the y-intercept.	
4 Sketch the graph.	

SB

p. 138

MATCHED EXAMPLE 7 Sketching $f(x) = x^{-1} = c$

Sketch the graph of $f(x) = \dfrac{1}{x} - 3$.

Steps	Working
1 Write the comparison with $y = x^{-1}$.	
2 State the asymptotes and vertices.	
3 Find the zero if possible.	
4 Sketch the graph.	

Show the horizontal asymptote as a dashed line, with its equation.

MATCHED EXAMPLE 8 | Sketching $f(x) = (x + b)^{-1}$

Sketch the graph of $f(x) = \dfrac{1}{x+2}$.

SB
p. 139

Steps	Working
1 Write the comparison with $y = x^{-1}$.	
2 State the asymptotes.	
3 State the y-intercept.	
4 Sketch the graph.	

Show the vertical asymptote as a dashed line, with its equation.

4

MATCHED EXAMPLE 9 | Sketching $f(x) = ax^{-1}$

Sketch the graph of $f(x) = \dfrac{2}{x}$.

Steps	Working
1 Write the comparison with $y = x^{-1}$.	
2 State the asymptotes.	
3 State the vertices.	
4 Sketch the graph.	

MATCHED EXAMPLE 10 | Sketching $f(x) = ax^{-2}$

SB

p. 140

Sketch the graph of $f(x) = \dfrac{1}{4x^2}$.

Steps	Working
1 Write the comparison with $y = x^{-2}$.	
2 State the asymptotes.	
3 State other reference points.	
4 Sketch the graph.	

MATCHED EXAMPLE 11 | Sketching a two-parameter truncus

Sketch the graph of $f(x) = \dfrac{4}{(x-2)^2}$.

Steps	Working
1 Write the comparison with $y = x^{-2}$.	
2 State the asymptotes.	
3 State other reference points.	
4 Sketch the graph.	

MATCHED EXAMPLE 12	Sketching a square root function

Sketch the graph of $f(x) = \sqrt{x-3}$.

Steps	Working
1 Compare to $y = \sqrt{x}$.	
2 State the starting point.	
3 State another point.	
4 Sketch the graph.	

4

MATCHED EXAMPLE 13 Sketching a cube root function

Sketch the graph of $f(x) = \sqrt[3]{x} + 3$.

Steps	Working
1 Compare to $y = \sqrt[3]{x}$.	
2 State the central point.	
3 Find the zero.	
4 Sketch the graph.	

MATCHED EXAMPLE 14 | Dilations

a The function $f(x) = x^2$ is dilated from the x-axis by the factor $\frac{1}{4}$ to give the function $g(x)$. State the rule for $g(x)$ and sketch f and g on the same axes.

b The function $f(x) = \sqrt[3]{x}$ is dilated from the y-axis by the factor 2 to give the function $g(x)$. State the rule for $g(x)$ and sketch f and g on the same axes.

SB

p. 145

Steps	Working
a **1** Use $g(x) = af(x)$.	
2 State the centres.	
3 State other points.	
4 Sketch the functions on the same axes.	
b **1** Use $g(x) = f(nx)$. $\frac{1}{n} = 2$, so $n = \frac{1}{2}$. $g(x) = f\left(\frac{1}{2}x\right)$	
2 State the starting points.	
3 State other points.	
4 Sketch the functions on the same axes.	

MATCHED EXAMPLE 15 | Reflections

a The function $f(x) = \sqrt[3]{x}$ is reflected in the x-axis to give the function $g(x)$.

State the rule for $g(x)$ and sketch f and g on the same axes.

b The function $f(x) = \dfrac{1}{x} - 4$ is reflected in the y-axis to give the function $g(x)$.

State the rule for $g(x)$ and sketch f and g on the same axes.

Steps	Working
a **1** Use $g(x) = -f(x)$.	
2 State the central points.	
3 State other points.	
4 Sketch the functions on the same axes.	
b **1** Use $g(x) = f(-x)$.	
2 State the asymptotes.	
3 State other points.	
4 Sketch the functions on the same axes.	

SB

p. 147

The function $f(x) = x^2$ is translated 4 units to the left and 2 units down to give the function $g(x)$. State the rule for $g(x)$ and sketch f and g on the same axes.

Steps	Working
1 Use $g(x) = f(x + b) + c$.	
2 State the turning points.	
3 State other points.	
4 Sketch the functions on the same axes.	

MATCHED EXAMPLE 17 | Combined transformations of points

The points $M(-5, -1)$, $N(6, -7)$ and $P(-2, 3)$ are on the function $f(x)$.

a Find the new points after a dilation by factor 4 in the x direction and by factor 2 in the y direction and translations 3 units in the negative y direction and 1 unit in the positive x direction.

b Find the positions of M, N and P after translations of 2 units up and 3 units to the right and then dilation by factor 2 from the x-axis and reflection in the y-axis.

Steps	Working
a 1 Write the transformation.	
2 Find the points.	
3 Write the answer.	
b 1 Write the transformation.	
2 Find the points.	
3 Write the answer.	

Sketch the graph of   **a** $f(x) = \dfrac{1}{2x-1} + 4$ **b** $g(x) = 2\sqrt{2x-6} + 2$

Steps	Working
a **1** State the basic function.	
2 State the transformations.	
Transformations parallel to the x-axis first.	
3 State the asymptotes.	
4 State the y-intercept.	
5 Find zeros if possible.	
6 Sketch the graph, including coordinates of important points and asymptote equations.	
b **1** State the basic function.	
2 Write the function in a suitable form.	
3 State the transformations.	
4 State the starting point.	
5 State other points.	

6 Sketch the graph.

MATCHED EXAMPLE 19 | Finding the equation of a transformed function

SB

p. 150

Each function $f(x)$ is transformed to $g(x)$ by the operations given. Find each function $g(x)$.

a $f(x) = x^2$: dilation from the x-axis by the factor 4, reflection in the y-axis and translation 2 down and 3 to the left.

b $f(x) = \sqrt{x+1}$: reflection in the x-axis, translation 3 in the positive direction of the x-axis and 5 in the negative direction of the y-axis and dilation by the factor $\dfrac{1}{2}$ from the y-axis.

Note that this is not in DRT order.

Steps	Working
a 1 Do the reflection in the y-axis.	
2 Do the translation to the left.	
3 Do the translation 2 down.	
4 Do dilation by a factor of 4 from the x-axis and write $g(x)$.	
b 1 Do the dilation using $\dfrac{1}{m}$.	
2 Do the translation to the right.	
3 Do the reflection in the x-axis.	
4 Do the translation down.	
5 Write g (x).	

| MATCHED EXAMPLE 20 | Finding the transformations from an equation |

What transformations have been performed on $f(x)$ to obtain $g(x)$?

a $f(x) = \sqrt{x}$, $g(x) = \sqrt{-4x-8} + 6$

b $f(x) = (2x+1)^2$, $g(x) = \dfrac{(2x-7)^2}{4} - 2$

Steps	Working

a **1** Express $g(x)$ in terms of f.

   2 Write as a series of transformations.

   3 Write in the correct order.

b **1** Express $g(x)$ in terms of f.

   2 Write as a series of transformations.

   4 Write in the correct order.

MATCHED EXAMPLE 21 | Finding transformations from a graph

This graph has been obtained by applying transformations to the graph of $y = x^2$.
What transformations have been used and what is the equation of the graph?

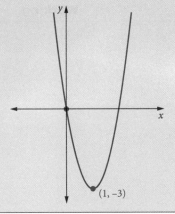

$(1, -3)$

Steps	Working
1 Use the turning point to find the translations.	
2 State the form of the equation.	
3 Substitute $(0, 0)$ to find the value of a and the equation.	
4 State the answer.	

SB

Using CAS 2:
Transformations
of power functions
p. 152

MATCHED EXAMPLE 22 | Existence of an inverse function

Determine which of the following functions are one-to-one and have inverse functions.

$f: R \to R, f(x) = x^2 + 1, g: R^+ \to R^+, g(x) = \sqrt{5x}, h: R \to R, h(x) = (x-1)^3$

Steps	Working
1 Consider $f(x)$.	
2 Consider $g(x)$.	
3 Consider $h(x)$.	
4 Write the answer.	

SB

p. 157

a Find the rule for the inverse of $f : R \to R, f(x) = \dfrac{1}{2x+3} - 1.$

b Is the inverse a function?

Steps	Working
a **1** Write with the variable y.	
2 Solve for x.	
3 Swap the variables to get the inverse relation.	
b Determine if it is a function.	

4

p. 158

MATCHED EXAMPLE 24 | The graph of a function and inverse relation

a Find the rule for the inverse of $y = \dfrac{1}{x+3} - 2$.

b Sketch the graph of $y = x$, the function and its inverse on the same axes.

c Comment on the graphs.

d State the domain and range of the function and its inverse.

e Write the inverse formally as a function if possible.

Steps	Working
a 1 Write with the variables swapped. **2** Solve for y.	
b Sketch the graphs on the same axes. $f(x)$ is defined only for $x \neq -3$, so the inverse relation is defined only for $x \neq -2$.	
c Comment on the graphs.	
d 1 State the domain and range of $f(x)$. **2** State the domain and range of the inverse.	
e The inverse is a function but does not have its maximal domain.	

9780170464093

MATCHED EXAMPLE 25 | Sketching the inverse from the graph

The graph of a function $f(x)$ is shown.

a Sketch the graph of the inverse of $f(x)$.

b Is the inverse a function?

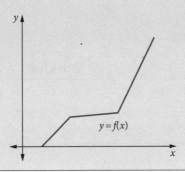

Steps	Working
a Reflect the graph in the line $y = x$.	
b Determine if the inverse (dashed) is a function.	

MATCHED EXAMPLE 26 | Matrix dimensions and elements

$$M = \begin{bmatrix} 1 & -2 & 0 & 3 & 2 \\ 1 & 5 & 2 & 3 & 0 \\ -2 & 3 & 2 & 3 & 0 \end{bmatrix}$$

a What are the dimensions of M?

b What are m_{13} and m_{24}?

Steps	Working
a Write the number of rows and columns.	
b m_{13} means row 1, column 3. m_{24} means row 2, column 4.	

$$M = \begin{bmatrix} 1 & -2 & 0 & 3 & 2 \\ 1 & 5 & 2 & 3 & 0 \end{bmatrix}$$

$$A = \begin{bmatrix} 1 & -2 \\ 3 & 0 \\ 5 & -1 \end{bmatrix} \text{ and } B = \begin{bmatrix} 3 & 0 \\ 4 & 1 \\ -2 & 2 \end{bmatrix}.$$

Find $4A$, $A + B$, $-A$, $A - B$ and $3A + 2B$.

Steps	Working
1 Multiply each element in A by 4.	
2 Add the corresponding elements.	
3 Multiply the elements of A by -1.	
4 Subtract the corresponding elements.	
5 Add $3A$ to $2B$.	

$$P = \begin{bmatrix} 1 & -1 \\ -2 & 0 \\ 3 & 2 \end{bmatrix} \text{ and } Q = \begin{bmatrix} 1 & 2 & 3 \\ 0 & -1 & -2 \end{bmatrix}.$$

Find PQ and QP.

Steps	Working
1 Write the product PQ.	
2 Multiply the numbers in each row of P by the numbers in each column of Q and add the answers.	
3 Write the product QP.	
4 Multiply the numbers in each row of Q by the numbers in each column of P and add the answers.	

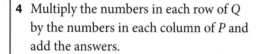

MATCHED EXAMPLE 29 | Matrix translations

Use matrices to find the image of the point of (0, 1) and the curve $y = 2x^3 + 1$ under the translation 3 to the right and 2 down.

Steps	Working
1 Write the translation as a transformation.	
2 Apply the transformation to the point (0, 1).	
3 Write the image point.	
4 Apply the transformation to a general point (x, y) of the curve.	
5 Solve the matrix equation for x and y.	
6 Substitute in the equation of the curve.	
7 Solve for y' to obtain the image of the function.	
8 Write the answer.	

4

MATCHED EXAMPLE 30 | Matrix dilations

Use matrices to find the image of the curve $y = \sqrt{x}$ under the dilation by a factor of 4 from the y-axis and $\dfrac{1}{4}$ from the x-axis.

Steps	Working
1 Write the dilation as a transformation.	
2 Apply the transformation to a general point (x, y) of the curve.	
3 Solve the matrix equation for x and y.	
4 Substitute in the equation of the curve.	
5 Solve for y' to obtain the image of the function.	
6 Write the answer.	

MATCHED EXAMPLE 31 | Matrix reflection

Use matrices to find the curve to which $y = x^3 + 4$ is mapped by reflection in the x-axis.

p. 170

Steps	Working
1 Write the reflection as a transformation.	
2 Apply the transformation to a general point (x, y) of the curve.	
3 Solve the matrix equation for x and y.	
4 Substitute in the equation of the curve.	
5 Write the answer.	

MATCHED EXAMPLE 32 | Combined matrix transformations

a What is the transformation for translation 3 in the negative x-direction and 2 in the positive y-direction, reflection in the y-axis and dilation from the x-axis by a factor of 2?

b Apply the transformation to find what the curve with equation $y = x^2 + 1$ is mapped to under the transformation.

Steps	Working
a 1 The translation is first.	
2 The reflection is second.	
3 Apply the last transformation. Notice the final order is *right-to-left*.	
4 Simplify.	
b 1 Apply the transformation to a general point (x, y) of the curve.	
2 Express x and y in terms of x' and y'.	
3 Substitute in the equation of the curve.	
4 Simplify to find y' for the image of the function.	
5 Write the answer.	

PROBABILITY AND COUNTING

CHAPTER 5

MATCHED EXAMPLE 1 | Finding the experimental and theoretical probabilities

p. 184

A coin is tossed 30 times and the following results are observed.

H, T, T, T, H, H, T, H, T, T, T, H, H, H, T, T, H, T, H, T, H, T, H, T, T, T, H, H, T, T

a Use the results obtained to find the experimental probability of the coin landing tails.

b Find the theoretical probability of the coin landing tails.

Steps	Working
a 1 Count the number of tails obtained in the 30 tosses of the coin.	
2 Find the probability by substituting into the experimental probability formula.	
b 1 List the sample space when a coin is tossed.	
2 Find the probability by substituting into the theoretical probability formula.	

MATCHED EXAMPLE 2 | Finding simple event probabilities

A number is randomly selected from the set {16, 17, 18, 18, 19, 20, 20, 21, 22, 23, 24, 25}

Let the random variable X = the number selected from the set.

Find

a $\Pr(X = 18)$ **b** $\Pr(X < 21)$

Steps	Working
a Use the theoretical probability formula. $$\Pr(E) = \frac{\text{number of favourable outcomes for event } E}{\text{number of possible outcomes}}$$	
b Count the number of digits in the sample space that are less than 21.	

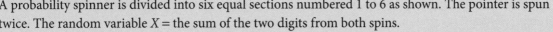

MATCHED EXAMPLE 3	Finding compound event probabilities	SB p. 188

A probability spinner is divided into six equal sections numbered 1 to 6 as shown. The pointer is spun twice. The random variable X = the sum of the two digits from both spins.

a Illustrate the sample space using a lattice diagram.

b Find

  i $\Pr(X = 7)$

  ii $\Pr(X \geq 7)$

5

Steps	Working
a Draw a 6 × 6 grid. Label the columns 1 to 6 for the first spin and the rows 1 to 6 for the second spin.	
b **i** Colour every cell where the first and second spins add to 7. There are 6 combinations that add to 7. (1,6) (2,5) (3,4) (4,3) (5,2) (6,1) There are 6 × 6 = 36 possible combinations. Find the theoretical probability.	
ii Colour all the combinations in the lattice that add to 7 or more. There are 21 combinations that add to 7 or more. Find the theoretical probability.	

SB

Using CAS 1:
Random integers
p. 191

p. 191

MATCHED EXAMPLE 4 | Using simulation to calculate probabilities.

There are 20 students in a year 11 class, and each day one student is selected at random each day to act as classroom monitor. The random numbers below simulate samples for 10 weeks.

10, 17, 16, 8, 1, 8, 6, 7, 14, 13, 12, 14, 4, 17, 7, 10, 9, 15, 13, 13, 14, 8, 8, 11, 13, 11, 17, 11, 14, 18, 2, 7, 17, 14, 16, 14, 15, 1, 18, 5, 17, 8, 1, 8, 4, 3, 8, 17, 18, 7

These random numbers can be grouped in sets of 5 to represent the students selected for 10 weeks.

Find, using this simulation, the probability that in a particular week a student is allocated classroom monitor duties more than once.

Steps	Working
1 Each student receives a number between 1 and 20. Group the random numbers into 10 groups of 5.	
2 Identify the groups that contain two identical numbers. These groups are weeks where a student has monitored class on more than one day in a week.	

MATCHED EXAMPLE 5 | Probabilities for non-equally likely outcomes using simulation

On average, a cricket player hits a boundary twice every 7 times at bat, and suppose he gets exactly two 'at bats' in every game. Using simulation, estimate the likelihood that the player will hit 2 boundaries in a single game.

Steps	Working
1 The two possible outcomes on each trial are either the player hits a boundary or the player does not hit a boundary. State which 2 of the 7 random numbers are a boundary and which are not a boundary.	
2 Generate data for 40 games with two innings per game, which is 80 random integers between 0 and 6. Pair off the random numbers so that each pair represents the two innings of a game.	
3 Two boundaries in a game are represented by a pair of numbers where both digits are either 4 or 6 or both.	
4 Calculated the estimated probability from the simulation.	

5

MATCHED EXAMPLE 6 The addition principle

Atlee needs to pick a book at random from a rack containing 5 Adventure books, 4 Horror books and 1 Motivational book. How many choices does Atlee have?

Steps	Working
1 Atlee can choose only one book in each group of books and in each case the word OR can be used between each of the choices. The word OR between the choices indicates the addition principle is used.	
2 Calculate the possibilities by adding the different options.	

MATCHED EXAMPLE 7	The multiplication principle

Ash is in a gift wrapping department and has 8 kinds of wrapping papers and seven colours of ribbons. How many different gift wrap styles are possible?

Steps	Working
1 Ash must choose one wrapping paper from eight wrapping papers AND one ribbon from seven colours of ribbon. This indicates the multiplication principle is used.	

5

MATCHED EXAMPLE 8 The addition and multiplication principles

Thomas visits a Chinese restaurant for lunch. He will either order a burger or noodles. The menu has burger with a choice of 2 different burger buns and 8 different kinds of patties For noodles there is a choice of 3 different noodles with 4 different sauces.

a How many different choices of burger are possible?

b How many different choices of noodles are possible?

c How many different choices does Thomas have altogether?

Steps	Working
a For burger, there are 2 different burger buns AND 8 different kinds of patties. AND is the multiplication rule.	
b For noodles, there are 3 different noodles AND 4 different sauces. AND is the multiplication rule.	
c Find the total number of choices. Thomas can have burger OR noodles. OR is the addition rule.	

Helen is in a birthday party of her friend Natasha, at the party guests can choose from a hotdog or hamburger as the main food, nachos, pretzels or fries as snacks, and ice cream or coke for dessert. Natasha guesses Helen will choose hamburger, fries and coke. Find the probability that Natasha is correct?

p. 197

Steps	Working
1 Find the total number of possible combinations.	
Helen is choosing from 2 main foods AND 3 snacks AND 2 desserts.	
2 Determine the probability.	
Helen has chosen one combination from 12 possible combinations.	

5

MATCHED EXAMPLE 10 | Permutations

A shopkeeper has to arrange four toys from a bag of nine toys in a rack. In how many ways can this be done?

Steps	Working
1 There are four places to fill, so draw four boxes.	
2 There are nine ways to fill the first position.	
3 Once the first place is filled, there are eight toys left to fill the second position.	
4 Once the second position has been filled, there are seven toys left to put in the third position.	
5 Once the third position has been filled, there are six toys left to put in the fourth position.	
6 Use the multiplication principle to calculate the number of possible arrangements.	

MATCHED EXAMPLE 11 | Factorials

In how many ways can the letters in the word FACTOR be arranged?

Steps	Working
1 We are required to arrange 6 letters of the word FACTOR. This can be done in 6! ways.	

SB

p. 200

SB

Using CAS 3:
Factorials p. 201

MATCHED EXAMPLE 12	Calculating permutations

Use the permutation formula to find the value of

a $^8P_3$	**b** $^6P_4$

Steps	Working
a Substitute $n = 8$ and $r = 3$ into the formula $^nP_r = \dfrac{n!}{(n-r)!}$	
b Substitute into the formula and calculate as the product of 4 consecutive descending numbers starting with 6.	

MATCHED EXAMPLE 13 | Arrangements with restrictions

Consider the five-letter arrangement of the word SOUTHERN. How many arrangements start with S and end with N?

Steps	Working
1 There are five places to arrange, so draw five boxes.	
2 The restriction requires the first letter be S and the fifth letter be N; therefore, the first and last positions are to be filled first. The first and last positions can be filled in one way each.	
3 The remaining three positions can now be arranged. There are six ways to fill the second position, five ways to fill the third position and four ways to fill the fourth position.	
4 Use the multiplication principle to calculate the number of arrangements.	

5

MATCHED EXAMPLE 14 | Arrangements with groups

In how many different ways can the letters of the word WRIGHT be arranged such that the letters G and H are always together?

Steps	Working
1 Put G and H in a group and count as one object or letter. $\widehat{GH}$WRIT There are five objects to arrange.	
2 Find the number of ways of arranging the letters in the group.	
3 Find the total number of possible arrangements by applying the multiplication principle.	

MATCHED EXAMPLE 15 | Probability involving arrangements

A 4-digit number is made from the digits {1, 2, 3, 4, 5, 6} where repeated digits are not allowed. Find the probability the number is an even number greater than 6000.

Steps	Working
1 Determine the number of possible 4-digit numbers if there is no restriction using the ^nP_r formula.	
2 Determine the number of 4-digit even numbers greater than 6000.	
There are 4 spaces to fill and the first and last places must be filled first.	
There is only one number {6} that can be placed first and two numbers {2, 4} that can be placed last.	
3 Fill the remaining positions with the 4 digits left.	
There are 4 numbers that can be placed second, leaving 3 numbers that can be placed third.	
4 Use the multiplication principle.	
5 Calculate the probability.	

MATCHED EXAMPLE 16 | Calculating combinations

Evaluate $^5C_2$.

Steps	Working
1 In the combination, the numerator is $^5P_2$, which is the product of 2 descending numbers starting with 5. The denominator is 2! **2** Cancel a factor of 2 and evaluate.	

MATCHED EXAMPLE 17 | Calculating combinations by symmetry

Evaluate $^8C_6$.

Steps	Working
By symmetry, $^8C_6 = {}^8C_2$.	
Evaluate $^8C_2$.	

p. 207

5

Using CAS 4:
Combinations
p. 207

MATCHED EXAMPLE 18 | Applying combinations

In a group of eight people, three prizes will be given. In how many ways can the prizes be distributed?

Steps	Working
Distribution of three prizes to three people is a combination because the order of distribution does not change the person.	
We are choosing three people from eight, which can be done in $^8C_3$ ways.	

9780170464093

SB

p. 207

A bag contains nine red marbles numbered 1 to 9 and seven green marbles numbered 1 to 7. In how many ways can three red and four green marbles be selected?

Steps	Working
1 The selection of number of different marbles is a combination. We are choosing three red marbles from nine red marbles AND four green marbles from seven green marbles.	
2 Use the multiplication rule to find the total number of ways.	

5

MATCHED EXAMPLE 20 | Finding probabilities involving combinations

A box contains 20 balloons of which 12 are blue and 8 are white. At random, 10 balloons are selected without replacement. Find the probability that of those 10 balloons 7 are blue.

Steps	Working
1 Count the number of ways of selecting 10 balloons where 7 are blue. In the sample, we must choose 7 blue balloons from a box containing 12 blue balloons AND choose 3 white balloons from a total of 8 white balloons.	
2 Determine the number of ways of selecting 10 balloons from a box of 20 balloons.	
3 Determine the probability.	

MATCHED EXAMPLE 21 | Finding the probability using counting methods

From seven men and six women, four are to be selected by Chris to form a committee.

a Find the probability he selects at least one woman.

b Chris lists the four names selected. Find the probability that the first and third names are of men.

Steps	Working
a 1 This is a combination since Chris is CHOOSING 4 persons from 13 persons. For at least one woman, Chris would need to choose 1, 2, 3 or 4 women. It is easier to find the number of combinations where no women are selected and then subtract the probability from 1. **2** Determine the number of combination if there is no restriction. **3** Determine the probability of at least one woman using the rule. Pr(at least 1) = 1 − Pr(none)	**a**
b 1 This is a permutation since Chris is ARRANGING four names. Count the number of arrangements where the first and third names are of men. There are 7 men that can go in first place and 6 remaining men that can go in third place. This leaves 11 persons for the second position and 10 persons for the fourth position. **2** Determine the number of permutations if there is no restriction. **3** Determine the probability.	**b**

SB

p. 222

MATCHED EXAMPLE 1 | Constant rate of change

For each graph, calculate, where possible, the constant rate of change over the entire interval, also stating whether the rate of change is positive, zero or negative. If not possible, give a reason.

a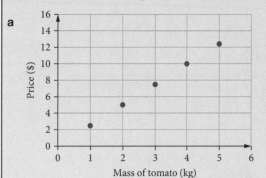

b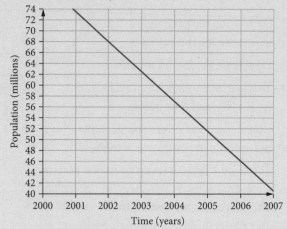

c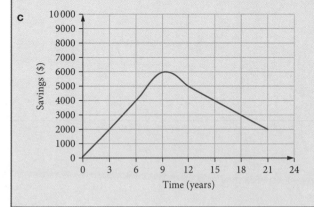

Steps	Working

a 1 Decide if the gradient is constant everywhere on the graph.

  2 Calculate the gradient.

    Choose two points on the line, say (2, 5) and (4, 10), and calculate the gradient of the line.

  3 State the rate of change.

b 1 Decide if the gradient is constant everywhere
on the graph.

2 Use the points (2002, 68) and (2006, 46) to
calculate the gradient of the line.

3 State the rate of change.

c 1 Decide if the gradient is constant everywhere
on the graph.

2 State the rate of change.

SB

p. 223

MATCHED EXAMPLE 2 | Rate of change from a linear graph

The table shows the weight, to the nearest kilogram, of a puppy over a 6-month period.

Month	5	6	7	8	9	10
Weight (kg)	8.1	9.3	10	10.7	11	12

a Construct a line graph of the data.

b Assuming the graph to be a linear function, describe the type of relationship that exists.

c Calculate the average growth rate of the puppy using the first and last points.

d Calculate the growth rate using the middle two points and compare it to your answer to **c**.

Steps	Working
a Plot and connect the points and decide whether the shape generally forms a straight line.	
b Describe the relationship.	
c Calculate the required gradient.	
d **1** Calculate the required gradient.	
2 Compare answers.	

MATCHED EXAMPLE 3 | Average rate of change

The position of a ball in the air is described by the equation $x = t^2 + 2t$, where x is in metres and t is in minutes. Find the average rate of change of the ball's position between $t = 2$ and $t = 4$.

Steps	Working
1 Work out the required x values.	
2 Write the coordinates.	
3 Calculate the gradient.	
4 State the average rate of change.	

SB

p. 228

MATCHED EXAMPLE 4 | Finding the tangent from a graph

This graph shows the radius of a balloon in centimetres as it is being inflated. Use a tangent to the graph to find the rate of change of radius of the balloon after 2 minutes.

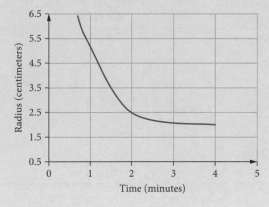

Steps	Working
1 Use a transparent ruler and pencil to draw a tangent at 2 minutes.	
2 Find two points on the tangent.	
3 Find the gradient of the tangent.	
4 Write the answer.	

SB

Using CAS 2:
Instantaneous rate
of change
p. 229

9780170464093

MATCHED EXAMPLE 5	Approximating the instantaneous rate of change

SB
p. 230

Let $y = x^2 + 4$.

a Determine approximations for the instantaneous rate of change of y with respect to x between $x = 2$ and $x = 2 + h$, using $h = 0.1, 0.01, 0.001$.

b What value will the approximation approach as the value of h approaches 0?

Steps	Working
a **1** Work out the required y values.	
2 Calculate each gradient. Use gradient $= \dfrac{y_2 - y_1}{x_2 - x_1}$	
b Look at the sequence of approximations.	

MATCHED EXAMPLE 6 | Approximating the rate of change from the right

Find the approximate rate of change of $f(x) = 4x^3 + x^2 - 4x + 1$ between the points with x-coordinates $x = 1$ and $x = 1.4$, approaching $x = 1$ from the right.

Steps	Working
1 Work out the value of h.	
2 Evaluate Δy.	
3 Calculate $\dfrac{\Delta y}{\Delta x}$.	
4 State the answer.	

MATCHED EXAMPLE 7 | Approximating the rate of change from the left

Use $h = 0.1$ to find the approximate rate of change of $f(x) = 4x^2 + 2x + 1$ at $x = 2$ by approaching $x = 2$ from the left.

Steps	Working
1 Find $x - h$.	
2 Calculate $f(x - h)$ and $f(x)$.	
3 Calculate $f(x) - f(x - h)$.	
4 Calculate $\dfrac{\Delta y}{\Delta x}$.	
5 State the answer.	

6

MATCHED EXAMPLE 8 | The derivative at a point

a Use $h = 0.1, 0.01, 0.001$ to find approximations for the instantaneous rate of change of $f(x) = x^4$ at $x = 2$ by approaching the point from the right.

b Hence, state the derivative of $f(x)$ at $x = 2$.

c Verify your answer using CAS.

Steps	Working
a Find each approximation using $$\frac{\Delta y}{\Delta x} = \frac{f(x+h) - f(x)}{h}$$	
b Deduce the derivative at the required value.	
c TI-Nspire ClassPad	

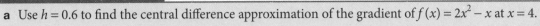

MATCHED EXAMPLE 9 | Central difference approximation

SB

p. 237

a Use $h = 0.6$ to find the central difference approximation of the gradient of $f(x) = 2x^2 - x$ at $x = 4$.

b Evaluate the gradient by approaching the function from the left and from the right.

c Compare the three methods, given that the exact value of the gradient is 15.

Steps	Working
a 1 Use the given value of h to evaluate $f(x+h)$ and $f(x-h)$.	
2 Apply the formula.	
b 1 To approach from the left, find $f(x)$, $f(x-h)$ and apply the formula.	
2 To approach from the right, find $f(x)$, $f(x+h)$ and apply the formula.	
c Compare each approximation to the exact answer.	

6

SB

p. 238

MATCHED EXAMPLE 10 | Finding the gradient function using approximations

Use $h = 0.2$ to find the gradient function $R(x)$ of $f(x) = 3x^2 - x$ for the x values 1, 2 and 3, approaching each x value from the right.

Steps	Working
1 Obtain $\dfrac{\Delta y}{\Delta x}$ for each x value given.	
2 Identify a relationship between the x value used and the gradient.	
3 State the gradient function, $R(x)$.	

SB

Using CAS 3:
The gradient
function
p. 239

Using CAS 4:
Stationary points
p. 240

Part of a truck's trip is represented by the graph.

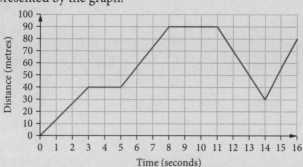

SB

p. 244

a How fast was the truck driven during the time interval 14 seconds to 16 seconds?

b For how long was the truck stationary?

c Find the speed during the return trip.

d Determine the truck's maximum speed.

Steps	Working
a **1** Calculate the speed in the required interval.	
2 State the answer, including units.	
b **1** Zero speed means the gradient is zero.	
2 State the answer.	
c **1** Calculate the speed in the required interval. Negative speed because the car is travelling towards 0.	
2 State the answer, including units.	
d **1** Maximum speed is the section with the steepest gradient.	
2 Calculate the speed in the required interval.	
3 State the answer, including units.	

p. 245

MATCHED EXAMPLE 12 | Interpreting a speed–time graph

The speed–time graph of a bike is shown.

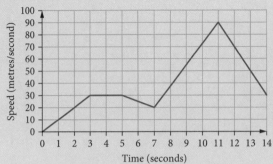

a State the acceleration of the bike between

   i 0 and 3 seconds

   ii 3 and 5 seconds

   iii 5 and 7 seconds

b In which time interval was the bike's acceleration greatest?

c Calculate the acceleration of the bike, in km/h^2, between 11 and 14 seconds.

Steps	Working
a **i** **1** Use $a = \dfrac{\Delta s}{\Delta t}$ between $t = 0$ and $t = 3$.	
2 State the answer, including units.	
ii **1** Use $a = \dfrac{\Delta s}{\Delta t}$ between $t = 3$ and $t = 5$.	
2 State the answer, including units.	
iii **1** Use $a = \dfrac{\Delta s}{\Delta t}$ between $t = 5$ and $t = 7$.	
2 State the answer, including units.	
b Consider the slope of the graph at each section.	
c **1** Find Δs and Δt between $t = 11$ and $t = 14$.	
2 Convert to km and hours.	
3 Use $a = \dfrac{\Delta s}{\Delta t}$.	
4 State the answer, including units.	

CHAPTER 7

THE EXPONENTIAL AND LOGARITHMIC FUNCTIONS

MATCHED EXAMPLE 1	Substituting and evaluating exponential functions

An exponential function is given by $f(x) = 4 \times 2^x$. Calculate the value of

a $f(1)$ **b** $f(0)$ **c** $f(-3)$

p. 258

Steps	Working
a Substitute $x = 1$ into the function and evaluate.	
b Substitute $x = 0$.	
c Substitute $x = -3$.	

MATCHED EXAMPLE 2	Simplifying exponential expressions
Simplify $2^{3x-2} \times 2^{4-2x}$.	

Steps	**Working**
Add powers and simplify.	

MATCHED EXAMPLE 3 | Sketching an exponential function $y = a^x$

Graph $y = 2^x$ for $[-3, 3]$ and comment on the graph's features.

Steps	Working
a 1 Draw up a table of values using the domain values given.	
2 Plot the points and join with a smooth curve.	
3 Examine the features of the graph.	

SB

Using CAS
1: Graphing
exponential
functions
p. 261

SB

p. 263

MATCHED EXAMPLE 4 | Sketching graphs of exponential functions

Sketch the graph of

a $y = 2(4^x) + 2$　　　　**b** $y = \dfrac{1}{4} \times 3^{3x}$

Steps	Working
a 1 Compare $y = 2(4^x) + 2$ to $y = Aa^{nx} + c$ and identify each transformation.	
2 Consider the new y-intercept and asymptote.	
3 Find an important point for scaling.	
4 Sketch the graph.	
b 1 Compare $y = \dfrac{1}{4} \times 3^{3x}$ to $y = Aa^{nx} + c$ and identify each transformation.	
2 Consider the new y-intercept and asymptote.	
3 Find an important point for scaling.	
4 Sketch the graph.	

MATCHED EXAMPLE 5 | Finding the rule of an exponential function

The rule for the function with the graph shown is of the form

$y = A(3^x) + c$

a Find the values of A and c.

b Hence, state the equation of the function.

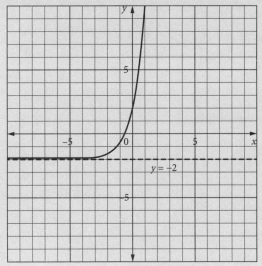

Steps	Working
a 1 Using the horizontal asymptote, find the value of c.	
2 Substitute $(0, 2)$ into the equation to find A.	
b Write the equation of the function.	

MATCHED EXAMPLE 6 | Exponential growth modelling

A colony of a single-cell organism is 5000 in number. The number of organisms doubles every day. Assuming that growth continues in this way, find how many organisms will be in the colony

a after 10 days **b** after 180 hours

Steps	Working
This is an example of exponential growth.	
The number of organisms doubles every day, or the new number of organisms is 200% of the previous number.	
The initial number is 5000.	
Write the growth function for the number of organisms after d days.	
a For the number of organisms after 10 days, substitute $d = 10$.	
b For the number of organisms after 180 hours, convert to days first. $$180 \text{ hours} = \frac{180}{24} = 7\frac{1}{2} \text{ days, so substitute } d = 7.5.$$	

MATCHED EXAMPLE 7 | Exponential decay modelling

The amount of water left in a leaking water tank is decreasing by 10% every minute.

When the leak first occurred, the amount of water in the tank was 550 litres. What will be the amount of water (correct to one decimal place) after 12 minutes?

Steps	Working
1 This is an example of exponential decay. The water decreases by 10% every minute, or the new water is 90% of the previous value. The initial value of water is 550 L.	
2 Write the decay function for water after m minutes.	
3 For the water amount after 12 minutes, substitute $m = 12$.	

7

SB

p. 268

MATCHED EXAMPLE 8 | Further exponential decay modelling

The difference between the temperature of an iron skillet and its surroundings decreases by 5% every minute. The skillet removed from the stove has a temperature of 120°C, and the room temperature is 20°C. Model the temperature of the iron skillet and find how long (to the nearest minute) it takes for the skillet to reach below 21°C.

Steps	Working
1 This is an example of exponential decay. The *difference* in temperature between the skillet and the room decreases by 5% each minute. Write the relationship between the functions. Room temperature = 20°C	
2 Write $D(m)$ as an exponential function. $D(m)$ decreases by 5% each minute, so $a = 95\% = 0.95$. The initial difference was 120°C − 20°C.	
3 Make $S(m)$ the subject.	
4 To find the time taken for the skillet to reach below 21°C, we want $S(m) < 21$.	
5 Try some different values of m until $C(m)$ is close to 21. OR use CAS to solve. **TI-Nspire** **ClassPad**	
6 Answer the question.	

MATCHED EXAMPLE 9 | Converting to logarithmic form

Write each statement in logarithmic form.

a $6^3 = 216$

b $4^{-2} = \dfrac{1}{16}$

Steps	Working
a Write as $\log_a (b) = x$, where a is the base and x is the power. Base $= 6$, power $= 3$	
b Base $= 4$, power $= -2$	

MATCHED EXAMPLE 10	Converting to exponential form

Write each statement in index form.

a $\log_4 (256) = 4$ **b** $\log_8 \left(\dfrac{1}{64}\right) = -2$

Steps	Working
a Write as $a^x = b$, where a is the base and x is the power. Base = 4, power = 4	
b Base = 8, power = −2	

MATCHED EXAMPLE 11	Evaluating logarithms

SB

p. 271

Evaluate each logarithm.

a $\log_5 (125)$ **b** $\log_7 (1)$ **c** $\log_3 \left(\dfrac{1}{81} \right)$

d $\log_2 (32)$ **e** $\log_{\frac{1}{6}} (36)$ **f** $\log_{\frac{1}{4}} \left(\dfrac{1}{16} \right)$

Steps	Working
a **1** Think: $5^? = 125$. **2** Evaluate the logarithm.	
b **1** Think $7^? = 1$. **2** Evaluate the logarithm.	
c **1** Think: $3^? = \dfrac{1}{81}$. The power must be negative. **2** Evaluate the logarithm.	
d **1** Think: $2^? = 32$. **2** Evaluate the logarithm.	
e **1** Think: $\left(\dfrac{1}{6} \right)^? = 36$. The power must be negative. **2** Evaluate the logarithm.	
f **1** $\left(\dfrac{1}{4} \right)^? = \dfrac{1}{16}$ **2** Evaluate the logarithm.	

7

MATCHED EXAMPLE 12	Evaluating logarithms using log laws

Evaluate each logarithm.

a $\log_2 (0.5)$ **b** $\log_1 (6)$ **c** $\log_2 (-2)$

d $\log_3 (0)$ **e** $\log_8 (8)$ **f** $\log_4 (1)$

Steps	Working
a The logarithm of the reciprocal of any number to its own base is -1 and $0.5 = \dfrac{1}{2}$.	
b The base of a log cannot be 1.	
c We can only have logarithms of positive numbers.	
d We can only have logarithms of positive numbers. You cannot find the logarithm of a negative number or 0. The base of a logarithm cannot be negative, 0 or 1.	
e The log of any number to its own base is 1.	
f The log of 1 to any base is 0.	

MATCHED EXAMPLE 13	Simplifying and evaluating logarithms using log laws

SB

p. 275

Simplify each expression.

a $\log_9 (3)$ **b** $\log_4 (32) + \log_4 (8)$ **c** $\dfrac{\log_4 (25)}{\log_4 (125)}$

Steps	Working
a Write 3 as a power of 9 and simplify.	
b 1 Use $\log_a (x) + \log_a (y) = \log_a (xy)$. **2** Write 256 as a power of 4 and simplify.	
c Write in terms of common powers of 5. Use $\log_a (x^n) = n \log_a (x)$ and simplify.	

7

SB

p. 275

MATCHED EXAMPLE 14 | Simplifying logarithms into a single expression

Simplify each expression to a single logarithm.

a $4\log_3(x) - 5\log_3(y+1) + 3\log_3(2x)$

b $5\log_2(2x) + 3\log_2\left(\dfrac{1}{x}\right)$

Steps	Working
a 1 Use $n\log_a(x) = \log_a(x^n)$.	
2 Use $\log_a(x) + \log_a(y) = \log_a(xy)$.	
3 Use $\log_a(x) - \log_a(y) = \log_a\left(\dfrac{x}{y}\right)$.	
b 1 Use $n\log_a(x) = \log_a(x^n)$.	
2 Use $\log_a(x) + \log_a(y) = \log_a(xy)$ and simplify.	

9780170464093

MATCHED EXAMPLE 15 | Evaluating logarithms using the change of base law

Use the change of base theorem to find the value of $\log_3(64)$ correct to three decimal places.

SB

p. 276

Steps	Working
1 Use the change of base theorem.	
2 Solve using CAS.	
	TI-Nspire **ClassPad**
3 Write the answer.	

MATCHED EXAMPLE 16 Graphing logarithmic functions

Sketch the graph of $f(x) = \log_4(x) + 2$, labelling important features.

Steps	Working
1 State the effect of the '+ 2'.	
2 Find the x-intercept by solving $f(x) = 0$.	
3 State the asymptote.	
4 Since there is no y intercept, state another point, or notice that the x-intercept $(1, 0)$ is translated up 2 to $(1, 2)$.	
5 Sketch the graph.	

SB

p. 280

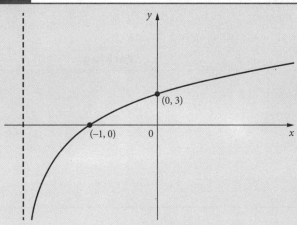

7

This is the graph of $y = a\log_2(x + b)$. Find the values of a and b and hence state the equation of the function.

Steps	Working
1 Substitute $(0, 3)$ into the equation.	
2 Substitute $(-1, 0)$ into the equation.	
3 Label the equations (1) and (2).	
4 Use (2) to solve for b.	
5 Substitute $b = 2$ into (1) to find a.	
6 Write the equation.	

MATCHED EXAMPLE 18 | Applying logarithmic functions

The New South Wales earthquake of 1989 had a seismographic reading of 3.184 m 800 km from its epicentre. What was its magnitude, correct to one decimal place?

Steps	Working
Substitute $x = 3184$ into the formula $(3.184 \text{ m} = 3184 \text{ mm})$.	
Write the answer.	

CHAPTER

PROBABILITY THEORY

8

SB

p. 291

MATCHED EXAMPLE 1 | Simple event probability

Alan has a fair die with six faces numbered 1 to 6. The probability of the die landing on one of these numbers on any of these throws is shown in the table below. Find the probability that Alan throws a number greater than 3.

Number, x	1	2	3	4	5	6
Probability, $\Pr(X = x)$	0.09	0.17	0.23	0.31	0.14	0.06

Steps	Working
1 The possible numbers that are greater than 3 are 4, 5 and 6.	
2 Use the values from the table.	

MATCHED EXAMPLE 2 | Finding probabilities using a Venn diagram 1

Out of 80 students, 60 play football, 45 play cricket and 30 play both football and cricket. Use a Venn diagram to find the probability of selecting a person who plays football but not cricket.

Steps	Working
1 Let *F* represent the set of students who play football and *C* represent the set of students who play cricket. Summarise the information given. $n(U)$ is the number of elements in the universal (total) set.	
2 Enter the probabilities in a Venn diagram. Fill the centre (intersection) first.	
3 In set *F*, there are 60 students. However, 30 have already been included in the intersection. The remaining part of set *F* contains $60 - 30 = 30$ students. Similarly, the remaining part of set *C* contains $45 - 30 = 15$ students. The total must be 80, leaving $80 - 30 - 15 - 30 = 5$ students outside sets *F* and *C*.	
4 The probability of selecting a person who plays football but not cricket is in the shaded section. 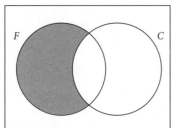	

MATCHED EXAMPLE 3 | Finding probabilities using a Venn diagram 2

If $Pr(A) = \dfrac{1}{4}$, $Pr(B) = \dfrac{2}{7}$ and $Pr(A \cap B) = \dfrac{1}{5}$, draw a Venn diagram and hence find $Pr(A' \cap B)$.

Steps	Working
1 Express all fractions with a common denominator.	
2 Illustrate the probabilities with a Venn diagram. Fill the centre (intersection) first.	
3 Complete the remaining part of set A and set B. $Pr(A' \cap B) = \dfrac{35}{140} - \dfrac{28}{140} = \dfrac{7}{140}$ $Pr(A' \cap B) = \dfrac{40}{140} - \dfrac{28}{140} = \dfrac{12}{140}$.	
4 All the probabilities must sum to 1 or $\dfrac{140}{140}$. Complete the remaining probability in the universal set. $Pr(A' \cap B') = \dfrac{40}{140} - \dfrac{7}{140} - \dfrac{28}{140} - \dfrac{12}{140} = \dfrac{93}{140}$.	
5 Find $Pr(A' \cap B)$, which is the part of set B, that is not in set A.	

MATCHED EXAMPLE 4 | Probabilities for mutually exclusive events

Events A and B are mutually exclusive with $\Pr(A) = \dfrac{3}{4}$ and $\Pr(B) = \dfrac{1}{8}$. Draw a Venn diagram and hence find $\Pr(A' \cap B')$.

Steps	**Working**
1 Express all fractions with a common denominator.	
2 Illustrate the probabilities with a Venn diagram. As A and B are mutually exclusive, the circles will not intersect.	
3 Enter the probabilities for A and B and calculate the remaining part of the universal set. $\Pr(A' \cap B') = \dfrac{8}{8} - \dfrac{1}{8} - \dfrac{6}{8} = \dfrac{1}{8}$	
4 Find the probability noting that $A' \cap B'$ is the region of the Venn diagram outside sets A and B.	

MATCHED EXAMPLE 5	Finding probabilities using two-way tables 1

SB

For two events A and B, $\Pr(A' \cap B) = 0.12$, $\Pr(A) = 0.65$ and $\Pr(B') = 0.45$. Find $\Pr(A \cap B)$.

p. 295

Steps	**Working**
1 Enter the given probabilities into a probability table.	
2 $\Pr(B)$ in the bottom row can be found first. $\Pr(B) + 0.45 = 1.0$ $\Pr(B) = 0.55$	
3 Complete the remainder of the table using the row totals and column totals.	
4 Find the value of $\Pr(A \cap B)$ in the table.	

9780170464093
Chapter 8 | Probability theory **159**

MATCHED EXAMPLE 6 | Finding probabilities using two-way tables 2

For two events A and B, $\Pr(A' \cap B') = 4p$, $\Pr(A \cap B') = p$, $\Pr(B) = 0.45$ and $\Pr(A) = 2p$. Find the value of p and hence find $\Pr(A \cap B')$.

Steps	Working
1 Enter the given probabilities into a probability table.	
2 $\Pr(B')$ can be found two ways. Adding the entries in the B' column $p + 4p = 5p$ and subtracting 0.45 from 1.0 $1.0 - 0.45 = 0.55$	
3 Substitute $p = 0.11$ and calculate the other values.	
4 Find the value of $\Pr(A' \cap B)$ in the table.	

MATCHED EXAMPLE 7 | Finding probabilities using two-way tables 3

For two events A and B, $\Pr(B) = 0.25$ and, $\Pr(A') = 0.18$. If A and B are independent, find $\Pr(A' \cap B')$.

Steps	Working
1 Enter the given probabilities into a probability table and complete the other totals.	
2 Use the rule for independent events to find $\Pr(A \cap B)$. $\Pr(A \cap B) = \Pr(A) \times \Pr(B)$	
3 Calculate the remaining probabilities in the table by subtracting from the totals. If A and B are independent then A and B' are also independent, $\Pr(A \cap B')$ can also be found using the formula $\Pr(A \cap B')$ $= \Pr(A) \times \Pr(B')$	
4 Find the value of $\Pr(A' \cap B')$ in the table.	

MATCHED EXAMPLE 8 | Using a tree diagram to find compound event probabilities

A bowl contains 60 marbles, of which 12 are blue in colour. Three marbles are randomly selected from the bowl, **with replacement.** This means that a marble is selected, its colour noted and the marble is replaced **before** the next marble is selected. Find the probability that 2 of the 3 marbles are not blue in colour.

Steps	**Working**
1 On each selection the marble selected can be blue (B) or not blue (B').	
2 Represent with a tree diagram.	
3 Identify the branches where there is one blue-coloured marble and two marbles that are not blue.	
4 Multiply the probabilities along the branches and add the products.	

MATCHED EXAMPLE 9 | Finding probabilities for independent events

In a fair, Anika tosses a ring on glass bottles that are aligned side by side. To win a prize, the ring should land on the neck of any bottle. Each of her tosses is independent and the probability of the ring landing on the neck of any bottle is 0.15. Find the probability (correct to three decimal places) that Anika's eighth toss wins her the prize.

Steps	**Working**
1 Write the sequence of events where the eighth shot lands on the neck of any bottle.	
2 Since each event is independent, we can multiply the probabilities.	
3 $\Pr(N) = 0.15$ and $\Pr(N') = 1 - 0.15 = 0.85$	

8

SB

p. 301

| **MATCHED EXAMPLE 10** | Finding probabilities for selections without replacement |

Sherin has six pairs of blue gloves and three pairs of green gloves. Her gloves are randomly mixed in her drawer. Sherin takes two individual gloves at random from the drawer in the dark.

Find the probability she selects a matching pair.

Steps	**Working**
1 There are 2 gloves selected without replacement from 12 blue gloves and 6 green gloves. Draw a tree diagram with two stages and two branches in each set.	
2 On the first selection, $\Pr(B) = \dfrac{12}{18}$, $\Pr(G) = \dfrac{6}{18}$	
3 If a blue glove is selected, there will be 11 blue and 6 green gloves. If a green glove is selected, there will be 12 blue and 5 green gloves.	
4 There will be 2 matching gloves if BB or GG is selected. Multiply probabilities along the branches and add between branches.	

SB

Using CAS 1:
Simulating
selections without
replacement
p. 302

For events A and B from a sample space, $\Pr(A' \cap B) = 0.15$, $\Pr(A) = 0.72$ and $\Pr(B') = 0.52$. Find $\Pr(A|B')$.

SB

p. 307

Steps	Working
1 Enter the given probabilities into a probability table and complete the other totals.	
2 Complete the probability table.	
3 Substitute into the conditional probability formula.	

8

SB

p. 307

MATCHED EXAMPLE 12 | Conditional probability using a Venn diagram

Market research on 80 people shows that 45 people drink coffee and 12 people drink either tea or coffee at breakfast. Find the probability of randomly selecting a person who drinks coffee given that a person who drinks tea at breakfast is selected.

Steps	**Working**

1 Express the information given using mathematical notation.

T represents tea drinkers and C represents coffee drinkers.

2 Enter and complete a Venn diagram.

$$Pr(C \cap T') = \frac{45}{80} - \frac{12}{80} = \frac{33}{80}$$

$$Pr(C' \cap T) = 1 - \frac{12}{80} - \frac{33}{80} = \frac{35}{80}$$

3 Write the conditional probability formula for this problem.

4 Calculate the probability.

We are only selecting from the 47 who drink tea and 12 of those who drink coffee.

It is also possible to substitute into the formula to obtain the same answer.

$$Pr(C|T) = Pr(C \cap T) \div Pr(T) = \frac{12}{80} \div \frac{47}{80}$$

$$Pr(C|T) = \frac{12}{80} \times \frac{80}{47}$$

$$Pr(C|T) = \frac{12}{47}$$

MATCHED EXAMPLE 13	Conditional probability from a tree diagram

A train in a small town that arrives every day at the station is on time 85% of the days when the weather in the town is sunny but only arrives on time 28% of the days when the weather is not sunny. The town's weather is sunny only 58% of the days. Find the probability that the train arrives at the station on time provided the weather is sunny.

SB p. 308

Steps	Working
1 Set up the tree diagram. *T* represents the train arrives on time. *T′* represents the train arrives late. *S* represents the weather is cloudy. *S′* represents the weather is not cloudy. The train arriving on time depends on the weather being sunny, so *S* and *S′* go first in the tree diagram.	
2 Enter the given probabilities on the tree diagram.	
3 Enter the remaining probabilities on the tree diagram. The probabilities on each set of branches must add to 1.0.	
4 Write the required probability using probability notation and express as a formula.	
5 Find the probability of $\Pr(T \cap S)$ and $\Pr(S)$ from the tree diagram.	
6 Substitute into the conditional probability formula.	

SB

p. 309

MATCHED EXAMPLE 14   Conditional probability using the law of total probability

A cricket team wins 52% of its games when it wins the toss and 12% of its games when the opposing team wins the toss. If the team wins the toss about 25% of the time, what fraction of the games does it win?

Steps	Working
1 Define variables.	
2 Write the probabilities for each of the defined events.	
3 Substitute into the law of total probabilities. $\Pr(A) = \Pr(A \mid B) \times \Pr(B) + \Pr(A \mid B') \times \Pr(B')$	

Michelle arranges flowers of a particular type in her vase every day. Her favourite flowers are lilies. If Michelle arranges lilies in her vase one day, there is a probability of 0.65 she will use other flowers the next day. If she does not use lilies on one day, then the probability of using lilies the next day is 0.75. On Monday, Michelle's vase is filled with lilies. Find the probability that Michelle arranges lilies in her vase twice in the next three days, correct to two decimal places.

SB
p. 310

8

Steps	Working
1 Set up the tree diagram. L represents arranging lilies. L' represents not arranging lilies.	
2 Enter the given probabilities on the tree diagram. Since Michelle arranges the vase with lilies on Monday, the probabilities for Tuesday are $\Pr(L) = 0.35$ $\Pr(L') = 0.65$	
3 Identify the branches where there are exactly two lilies.	
4 Calculate the probabilities.	

p. 315

MATCHED EXAMPLE 16 Finding the probability of the union of two events from a two-way table

If $\Pr(A) = 0.32$, $\Pr(B) = 0.45$ and $\Pr(A \cap B') = 0.16$, find $\Pr(A \cup B)$.

Steps	Working
1 Enter the probabilities into a probability table.	
2 Complete the table.	
3 Calculate $\Pr(A \cup B)$ using the addition rule.	

If $\Pr(A \cup B) = \dfrac{1}{8}$, $\Pr(A \cap B) = \dfrac{3}{5}$ and $\Pr(B) = 1 - 2\Pr(A)$, find $\Pr(B)$.

SB

p. 316

Steps	Working
1 Let $\Pr(A) = a$. Write $\Pr(B)$ in terms of a.	
2 Substitute the given values into the addition rule.	
3 Substitute into $\Pr(B) = 1 - 2a$.	

MATCHED EXAMPLE 18 | Finding Pr($A \cup B$)

A card is selected at random from a pack of 52 playing cards. Find the probability of selecting a black card or an ace.

Steps	Working
1 Pr(black or ace) = Pr($B \cup A$) The addition rule for probabilities can be used to solve this problem.	
2 Find the probabilities required in the formula. Pr($B \cap A$) = Pr(black and an ace)	
3 Substitute into the addition rule.	

CHAPTER 9

THE CIRCULAR FUNCTIONS

MATCHED EXAMPLE 1	Converting between degrees and radians

Convert

a $\dfrac{3\pi}{2}$ to degrees   **b** 120° to radians   **c** 210° to radians

SB
p. 328

Steps	Working
a Substitute $\pi = 180°$ and simplify.	
b 1 Start with $180° = \pi$ and divide to find 1°.	
2 Multiply to find 120° and simplify.	
c Multiply by $\dfrac{\pi}{180°}$	

SB

Using CAS 1:
Setting degree and
radian mode
p. 328

MATCHED EXAMPLE 2 | Finding unknown sides in right-angled triangles

Find, correct to three decimal places, the value of the pronumeral in each triangle.

a   **b**

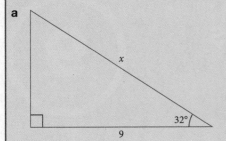

Steps	Working
a 1 Set the calculator to degree mode.	
2 Decide which trigonometric ratio is required. With 32°, x is the hypotenuse and 9 is the adjacent side, so use cos.	
3 Divide both sides by cos 32° and multiply both sides by x to solve the equation.	
4 Evaluate and round the answer.	
b 1 Set the calculator to radian mode.	
2 Decide which trigonometric ratio is required. With $\dfrac{2\pi}{5}$, y is the hypotenuse and 14 is the opposite side, so use sin.	
3 Multiply both sides by y.	
4 Divide both sides by $\sin\left(\dfrac{2\pi}{5}\right)$ to solve the equation.	

a Find, correct to one decimal place, angle θ in radians.

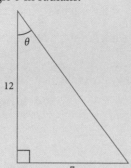

b Find to the nearest degree, the size of angle θ.

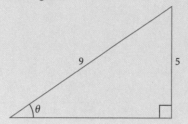

Steps	Working
a 1 Set the calculator to radian mode.	
2 Decide which trigonometric ratio is required. With θ, 12 is the adjacent side and 7 is the opposite side, so use tan.	
3 The angle is the inverse of the trigonometric function.	
4 Evaluate and round the answer.	
b 1 Set the calculator to degree mode.	
2 Decide which trigonometric ratio is required. With θ, 5 is the opposite side and 9 is the hypotenuse, so use sin.	
3 The angle is the inverse of the trigonometric function.	
4 Evaluate and round the answer.	

SB

p. 331

MATCHED EXAMPLE 4 | Arc length 1

a Find the exact length of the arc formed by an angle of $\dfrac{\pi}{4}$ in a circle of radius 12 cm.

b Find (correct to two decimal places) the angle subtended at the centre of a circle of radius 6.2 cm by an arc 2.8 cm long.

Subtended means 'sits opposite'.

c Find the exact length of the arc formed by an angle of 30° in a circle of radius 5 cm.

Steps	Working
a Use $l = r\theta$ with $r = 12$ cm and $\theta = \dfrac{\pi}{4}$.	
b **1** Use $l = r\theta$ with $l = 2.8$ cm, $r = 6.2$ cm. **2** Divide both sides by 6.2.	
c **1** Convert the angle into radians first. **2** Use $l = r\theta$ with $r = 5$ cm and $\theta = \dfrac{\pi}{6}$.	

9780170464093

MATCHED EXAMPLE 5 | Arc length 2

The length of a swing in a playground is 2 metre long. If a child on the swing travelled an arc of 2.5 metre, through what angle, to the nearest degree, did the swing rotate?

Steps	Working
1 Use $l = r\theta$ with $l = 2$ m, $r = 2.5$ m.	
2 Divide both sides by 8.	
3 Convert this angle from radians to degrees.	

MATCHED EXAMPLE 6 | Using the Pythagorean identity

Given $\sin(\theta) = -\dfrac{3}{7}$ and $\dfrac{3\pi}{2} < \theta < 2\pi$, find the exact value of

a $\cos(\theta)$ **b** $\tan(\theta)$

Steps	Working

a 1 Use the identity $\sin^2(\theta) + \cos^2(\theta) = 1$ and solve to find $\cos(\theta)$.

> This question can also be answered by first drawing a right-angled triangle with opposite side 3 and hypotenuse 7 and using Pythagoras' theorem to find the adjacent side.

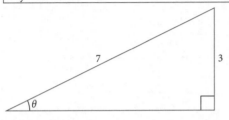

2 Use the quadrant to determine if $\cos(\theta)$ is positive or negative.

$\dfrac{3\pi}{2} < \theta < 2\pi$ means 4th quadrant.

b Use $\tan(\theta) = \dfrac{\sin(\theta)}{\cos(\theta)}$.

| MATCHED EXAMPLE 7 | Applying Pythagoras' theorem to find a trigonometric ratio |

SB

p. 338

Given $\cos(\theta) = \dfrac{3}{8}$ and $0° < \theta < 90°$, find the exact value of $\sin(\theta)$.

Steps	Working

1 Draw a right-angled triangle with adjacent side 3 and hypotenuse 8, and use Pythagoras' theorem to find the opposite side.

2 Find $\sin(\theta)$.

3 Use the quadrant to determine if $\sin(\theta)$ is positive or negative.

$0° < \theta < 90°$ means 1st quadrant.

p. 339

MATCHED EXAMPLE 8 | Equivalent ratios

State the equivalent trigonometric function in quadrant 1 for each of the following.

a $\sin(180° - 60°)$　　　**b** $\cos(360° - 20°)$　　　**c** $\tan\left(\dfrac{\pi}{10} - 2\pi\right)$

Steps	Working
a Use $\sin(180° - \theta°) = \sin(\theta°)$.	
b Use $\cos(360° - \theta°) = \cos(\theta°)$.	
c **1** Write in the form $\tan(2\pi - \theta)$.	
2 Use $\tan(-\theta) = -\tan(\theta)$.　Apply $\tan(2\pi - \theta) = -\tan(\theta)$.	

MATCHED EXAMPLE 9 | Rewrite the angle to find equivalent ratios

State the equivalent trigonometric function in quadrant 1 for each of the following.

a $\cos(210°)$ **b** $\sin\left(\dfrac{5\pi}{4}\right)$ **c** $\tan\left(\dfrac{5\pi}{3}\right)$

Steps	Working
a **1** Write as $\cos(\pi + \theta)$. **2** Use $\cos(\pi + \theta) = -\cos(\theta)$.	
b **1** Write as $\sin(\pi + \theta)$. **2** Use $\sin(\pi + \theta) = -\sin(\theta)$.	
c **1** Write in the form $\tan(2\pi - \theta)$. **2** Use $\tan(2\pi - \theta) = -\tan(\theta)$.	

9

| MATCHED EXAMPLE 10 | Exact values |

Find the exact value of each trigonometric ratio.

a $\sin(225°)$ **b** $\cos\left(\dfrac{7\pi}{3}\right)$ **c** $\tan\left(-\dfrac{4\pi}{3}\right)$

Steps	Working
a **1** Identify the quadrant of the angle and use ASTC to establish the sign. **2** For the 3rd quadrant, write the angle in the form $\pi + \theta$, using an acute angle θ.	
b **1** Identify the quadrant of the angle and use ASTC to establish the sign. **2** For the 1st quadrant, write the angle in the form $2\pi + \theta$, using an acute angle for θ.	
c **1** Use $\tan(-\theta) = -\tan(\theta)$. **2** Identify the quadrant of the angle and use ASTC to establish the sign. **3** For the 3rd quadrant, write the angle in the form $\pi + \theta$, using an acute angle for θ.	

MATCHED EXAMPLE 11 | Amplitude, period and mean

a State the amplitude, period and mean value of $y = 3 \sin (2x) - 2$.

b Sketch the graph of $y = 3 \sin (2x) - 2$ in $\left[0, \dfrac{5\pi}{4} \right]$.

Steps	Working
a $y = a \sin (nx) + c$ has amplitude a, period is $\dfrac{2\pi}{n}$ and mean value c.	
b 1 Decide on a scale for the x-axis.	
2 Decide on a range for the y-axis.	
3 Draw the grid and indicate the mean value, the maximum and the minimum.	
4 Find some critical points, including the minimum. Use the period to find other values.	

5 Plot and connect the points smoothly, keeping in mind the general shape of a sine graph.

Approximating sin (*x*)

Calculate the percentage error, correct to three decimal places, in the approximation to sin (*x*) using *x* = 0.2.

SB

p. 349

Steps	Working
1 Use sin (*x*) ≈ *x* to obtain the approximation at the given value.	
2 Calculate the exact value with a calculator.	
3 Calculate the error, the difference between the exact value and the approximation.	
4 Calculate the percentage error, the error as a percentage of the exact value.	

SB

Using CAS 2:
Graphing cosine
p. 351

MATCHED EXAMPLE 13 | Transformations applied to cos (x)

a State the transformations required for $y = \cos (x)$ to become $y = 2 \cos (3x) - 2$.

b Sketch each transformation in the interval $[0, 2\pi]$.

c State the domain and range of $y = 2 \cos (3x) - 2$ in the interval $[0, 2\pi]$.

Steps	Working
a Compare $y = 2 \cos (3x) - 2$ to $y = a \cos (nx) + c$ and identify each transformation.	
b Draw the graph of $y = \cos (x)$ with the graph of each transformation.	
c State the domain and range.	

MATCHED EXAMPLE 14 | Transformations applied to tan (x)

a State the transformations required to transform the graph of $y = \tan (x)$ to the graph given by $y = 2\tan (2x) + 1$ in the interval $[0, \pi]$.

b Find the equations of all asymptotes in the given interval.

c Sketch $y = 2\tan (2x) + 1$, including asymptotes.

d State the domain and range of $y = 2\tan (2x) + 1$ in the given interval.

Steps	**Working**
a Compare $y = 2\tan (2x) + 1$ to $y = a \tan (nx) + c$ and identify each transformation.	
b Asymptotes are $x = \dfrac{k\pi}{2n}$ for odd k.	
c Draw the graph of $y = \tan (x)$ with the graph of each transformation. $y = \tan (x)$ in grey, $y = \tan (2x)$ in dashed blue, $y = 2\tan (2x)$ in black, $y = 2\tan (2x) + 1$ in dark blue.	
d State the domain and range.	

p. 361

MATCHED EXAMPLE 15 Applying circular functions

The graph shows the vertical distance between a rider on a rotating Ferris wheel and the centre of the wheel, over time.

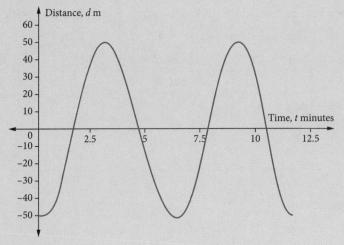

a What is the radius of the wheel?

b How long does it take the wheel to make one revolution?

c What is the minimum vertical distance reached by the rider? Where on the wheel is this?

d What do the t-intercepts on the graph represent?

Steps	Working
a Find the amplitude.	
b One revolution of the wheel corresponds to 1 period on the graph.	
c Find the lowest point on the graph.	
d The t-intercepts represent when $d = 0$, when the point is 0 m vertically from the centre of the wheel.	

MATCHED EXAMPLE 16 | Water level described by a trigonometric function

The water level on a pier in the ocean as it changes periodically due to the tides during a certain day and is described by the function $h = 4\cos\left(\dfrac{\pi}{3}t\right) + 10$, where h is in metre and t is the time in hours after midnight.

a Sketch the graph of the function in the interval [0, 15].

b At what times after midnight is the water level at its maximum?

c What is the water level at 2:30 a.m.? Give your answer correct to three decimal places.

Steps	Working
a Identify the amplitude, period and transformations, and use them to sketch the graph.	
b Find the t values after $t = 0$ corresponding to the maximum points on the graph.	
c Determine the t value corresponding to 2:30 a.m. and substitute to find h.	

SB

Using CAS 3:
Applying circular
functions
p. 363

CHAPTER

10 DIFFERENTIATION

MATCHED EXAMPLE 1	Limit of a function 1

Find the value of $\lim\limits_{x \to 3} f(x)$, $f(x) = \dfrac{9 - x^2}{3(x-3)}$.

Steps	Working
1 Simplify to obtain a linear function.	
2 Decide if the limit is the same from the left and right sides.	
3 Answer the question.	

MATCHED EXAMPLE 2 | Limit of a function 2

Evaluate $\lim\limits_{x \to -1} \dfrac{3}{1+x}$.

Steps	Working
1 Look at the limit from the left.	
2 Look at the limit from the right.	
3 Decide if the limit exists.	

Using CAS 1:
Limits
p. 379

SB

p. 380

MATCHED EXAMPLE 3 | Testing for a continuous function

Determine whether each function is continuous at the given value.

a $f(x) = \dfrac{x^2 - 9}{x + 3}$ at $x = 3$ **b** $f(x) = \sqrt{x - 5}$ at $x = 9$

Steps   **Working**

a **1** Check whether $f(3)$ exists.

2 Evaluate $\lim\limits_{x \to 3} f(x)$.
When x approaches 3 from the negative side
$(x \to 3^-)$, $\dfrac{x^2 - 9}{x + 3}$ becomes closer to 0.
When x approaches 3 from the positive side
$(x \to 3^+)$, $\dfrac{x^2 - 9}{x + 3}$ becomes closer to 0.

3 Check whether $\lim\limits_{x \to 3} f(x) = f(3)$.

4 Answer the question.

b **1** Check whether $f(9)$ exists.

2 Evaluate $\lim\limits_{x \to 9} f(x)$.
When x approaches 9 from the negative side
$(x \to 9^-)$, $\sqrt{x - 5}$ becomes closer to 2.

When x approaches 9 from the positive side
$(x \to 9^+)$, $\sqrt{x - 5}$ becomes closer to 2.

3 Check whether $\lim\limits_{x \to 9} f(x) = f(9)$.

4 Answer the question.

MATCHED EXAMPLE 4	Differentiation by first principles

Use differentiation by first principles to find the gradient function for $f(x) = 3x^3$.

Steps	Working
1 Write the function.	
2 Find an expression for $f(x+h)$. Expand and simplify.	
3 Find an expression for $\dfrac{f(x+h)-f(x)}{h}$ and simplify.	
4 Find $\lim\limits_{h \to 0} \dfrac{f(x+h)-f(x)}{h}$.	
5 State the gradient function.	

MATCHED EXAMPLE 5 | Derivative at a point by first principles

Use differentiation by first principles to find $f'(x)$ for $f(x) = 2x^2 - x$, then evaluate $f'(1)$.

Steps	Working
1 Write the function.	
2 Find an expression for $f(x + h)$. Expand and simplify.	
3 Find an expression for $\dfrac{f(x+h)-f(x)}{h}$ and simplify.	
4 Find $f'(x)$.	
5 Evaluate $f'(x)$ using the given value.	

MATCHED EXAMPLE 6 | Checking for differentiability

Decide if each function is differentiable at the given point.

a $f(x) = \dfrac{1}{x-4}$ at $x = 2$ **b** $f(x) = \dfrac{x(x-3)}{(x-3)}$ at $x = 3$

SB
p. 386

Steps	Working
a **1** Decide if the function is continuous at $x = 2$.	
2 Decide if the function has a unique tangent at $x = 2$.	
3 Decide if the function is differentiable.	
b **1** Simplify the function.	
2 Decide if the function is continuous.	
3 Decide if the function is differentiable.	

SB

p. 387

MATCHED EXAMPLE 7	Term-by-term differentiation

Differentiate each function.

a $f(x) = 4x^3 - 3x$ **b** $f(x) = 2x^5 - 0.25x^3 + 12x$

Steps	Working
a Differentiate using $f'(x) = a \times nx^{n-1}$.	
b Differentiate each term separately using $f'(x) = a \times nx^{n-1}$.	

Differentiate each function.

a $f(v) = 6v^{-\frac{2}{3}}$ **b** $f(r) = \dfrac{r^2}{3} + 2r^3$ **c** $y = -\sqrt{t^5}$

SB

p. 388

10

Steps	Working
a Differentiate with respect to v.	
b Differentiate with respect to r.	
c **1** Write the function in the form $y = at^n$. **2** Differentiate with respect to t.	

MATCHED EXAMPLE 9 The derivative at a given point

Calculate $f'(x)$ for the given value of x.

a $f(x) = 0.5x^4 + 2x^{\frac{1}{2}}$, $f'\left(\dfrac{1}{4}\right)$ **b** $f(x) = x^{-2} + 3x^2 + 4x$, $f'(1)$

Steps	Working
a 1 Differentiate each term.	
2 Evaluate $f'\left(\dfrac{1}{4}\right)$.	
b 1 Differentiate using $f'(x) = anx^{n-1}$.	
2 Evaluate $f'(1)$.	

MATCHED EXAMPLE 10	Finding a constant given the derivative at a given point

For the function $f(x) = 2x^3 - ax^2 + 1$, $f'(4) = 24$. Find the value of the constant, a.

Steps	Working
1 Find the derivative of $f(x)$.	
2 Form an equation.	
3 Solve for the unknown.	

SB
p. 390

SB

Using CAS 3:
Finding the
function from the
derivative
p. 390

MATCHED EXAMPLE 11 | Applying the central difference formula to find gradient

Use the central difference formula with $h = 0.1$ to find an estimate for the gradient at the point on the curve $f(x) = x^4$ when $x = 3$.

Steps	Working
1 Evaluate $f(a - h)$ and $f(a + h)$ using the known information.	
2 Substitute the values into the formula.	

MATCHED EXAMPLE 12 | Using the central difference formula to find velocity and acceleration

SB

p. 394

The position, $f(t)$ metres, of a particle at time t seconds is described by the equation $f(t) = 2t^2 - 3t + 2$.

a Use a value of $h = 0.01$ to estimate the particle's instantaneous velocity after

   i 15 s   **ii** 25 s

b Hence, estimate the particle's average acceleration from 15 s to 25 s.

Steps	Working
a **i** Use the central difference formula to estimate the gradient at $t = 15$ s.	
ii Use the central difference formula to estimate the gradient at $t = 25$ s.	
b Use the estimated values for the instantaneous velocity to obtain the average acceleration.	

Using CAS 4:
Finding
acceleration
using the central
difference
p. 395

MATCHED EXAMPLE 13 | Equation of the tangent at a given point

Find the equation of the tangent to the curve $f(x) = 2x^2 - 3x + 2$ at the point $(0, 2)$.

Steps	Working
1 Find the gradient function by differentiation.	
2 Find the gradient of the tangent by substituting $x = 0$ in $f'(x)$.	
3 Find the equation of the tangent using $y - y_1 = m(x - x_1)$. OR Find the equation of the tangent using $y = mx + c$.	

9780170464093

SB

p. 399

Find the point(s) on the graph of $f(x) = x^3 + 3x^2 - x + 2$ where the tangent:

a has a gradient of 8. **b** is parallel to the line $y = -x + 3$.

Steps	Working
a **1** Find the gradient function by differentiation.	
2 Solve $f'(x) = 8$.	
3 Find the corresponding y values.	
4 State the coordinates required.	
b **1** Determine the gradient value needed.	
2 Solve $f'(x) = -1$.	
3 Find the corresponding y values.	
4 State the coordinates required.	

CHAPTER

11 TRIGONOMETRIC AND EXPONENTIAL EQUATIONS

SB

p. 413

MATCHED EXAMPLE 1	Solving trigonometric equations.

Solve $\sin(x) = -\dfrac{1}{2}$, $x \in [0, 2\pi]$ for x.

Steps	Working

1 Ignoring the sign of $-\dfrac{1}{2}$, find the reference

(the acute angle in the first quadrant).

Refer to the exact value table.

2 Identify the quadrants where there are solutions.

Find the quadrants where $\sin(x)$ is negative.

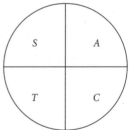

3 Find the solutions.

Find the equivalent angle to $\dfrac{\pi}{6}$ in quadrants 3 and 4.

Solve $\sin(2x) = \dfrac{\sqrt{3}}{2}$, $x \in [0, 2\pi]$ for x.

Steps	Working

1 Find the reference angle.

Refer to the exact value table.

2 Identify the quadrants where there are solutions.

Find the quadrants where sin (2*x*) is positive.

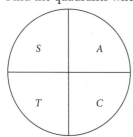

3 Determine the domain for 2*x*.

If $x \in [0, 2\pi]$, then $2x \in [0, 4\pi]$.

A domain of $[0, 4\pi]$ will produce 4 solutions.

4 Solve the equation for 2*x*.

$\dfrac{\pi}{3}$ and $\dfrac{2\pi}{3}$ are in the first revolution.

To obtain the next two solutions, add 2π or $\dfrac{6\pi}{3}$ to these values.

5 Divide by 2 to solve for *x*.

6 Confirm by CAS. Be sure to include the domain

TI-Nspire   ClassPad

MATCHED EXAMPLE 3 | Transposing and solving trigonometric equations

Solve $\sqrt{2}\sin(2x)+1=0$, $x \in [-\pi, 0]$ for x.

Steps	Working
1 Transpose the equation into the form $\sin(2x) = \ldots$	
2 Find the reference angle.	
3 Identify the quadrants where $\sin(2x)$ is negative. 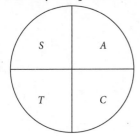	
4 Determine the domain for $2x$. If $x \in [-\pi, 0]$, then $2x \in [-2\pi, 0]$.	
5 Solve the equation for nx. For negative angles, start at 0 on the x-axis and rotate clockwise.	
6 Divide by 2 to solve for x.	

MATCHED EXAMPLE 4	Solving equations of the form sin (*nx*) = *a* cos (*nx*)

Solve $\sin(2x) = -\sqrt{3}\cos(2x)$, $0 \le x \le 2\pi$ for *x*.

Steps	**Working**
1 Divide both sides by cos (2*x*).	
2 Determine the reference angle.	
3 Identify the quadrants where tan (2*x*) is negative.	
4 Determine the domain for 2*x*.	
If $x \in [\,0, 2\pi]$, then $2x \in [\,0, 4\pi]$.	
5 Solve the equation for 2*x*.	
$\dfrac{2\pi}{3}$ and $\dfrac{5\pi}{3}$ are in the first revolution.	
For the second revolution, add 2π or $\dfrac{6\pi}{3}$ to these values.	
6 Divide by 2 to solve for *x*.	

p. 419

MATCHED EXAMPLE 5 | Solving simple exponential equations

Solve $25^x = 625$ for x.

Steps	Working
1 Write 25 and 625 as powers of 5.	
2 Equate the powers.	
3 Solve the linear equation.	

MATCHED EXAMPLE 6	Solving exponential equations

Solve $27^{x-3} = 9^{4-x}$ for x.

Steps	**Working**
1 Write 27 and 9 as powers of 3.	
2 Expand the brackets and multiply the powers.	
3 Equate the powers.	
4 Solve the linear equation.	

SB

p. 420

MATCHED EXAMPLE 7 | Solving exponential equations using index laws

Solve $8^{2x} \times 64^{3x-4} = 256^{5-x}$ for x.

Steps	Working
1 Write 8, 64 and 256 as powers of 2.	
2 Expand the brackets, multiply the powers and then simplify.	
3 Equate the powers and solve the linear equation.	
4 Confirm by CAS	

TI-Nspire

ClassPad

MATCHED EXAMPLE 8 | Solving exponential quadratic equations 1

Solve $2^{2x} - 12(2^x) + 32 = 0$ for x.

Steps	Working
1 Let a represent the exponential term.	
2 Express the equation in terms of a.	
3 Factorise and solve for a.	
4 Rewrite the solutions in terms of x.	
5 Solve by writing all terms as a power of 2 and equating the powers.	

11

MATCHED EXAMPLE 9 | Solving exponential quadratic equations 2

Solve $3(3^{2x}) - 4(3^x) + 1 = 0$ for x.

Steps	Working
1 Let k represent the exponential term.	
2 Express the equation as a quadratic in terms of k.	
3 Factorise and solve for k.	
4 Rewrite the solutions in terms of x.	
5 Solve by writing all terms as a power of 3 and equating the powers.	

MATCHED EXAMPLE 10 | Solving logarithmic equations

Solve $\log_4(2x-4)=1$ for x.

p. 423

Steps	Working
1 Change the equation from log form to index form.	
2 Simplify and solve the equation.	

MATCHED EXAMPLE 11 Solving logarithmic equations using logarithm laws

Solve $\log_3(2x-3)+3\log_3(2)=4$ for x.

Steps	Working
1 Use log laws to express the left-hand side of the equation as a single logarithm.	
2 Change the equation from log form to index form.	
3 Simplify and solve the equation.	

MATCHED EXAMPLE 12 | Solving equations where every term is a logarithm

Solve $\log_4(x) = 3\log_4(2) + \log_4(7)$ for x.

Steps	Working
1 Use log laws to express the right-hand side of the equation as a single logarithm.	
2 Equate the brackets.	

p. 424

MATCHED EXAMPLE 13 | Solving natural logarithmic equations 1

Solve $\log_e(3x-2)+\log_e(3x)=\log_e(24)$ for x.

Steps	Working
1 Use log laws to express the left-hand side of the equation as a single logarithm.	
2 Equate the brackets and solve the quadratic.	
3 Check to see if both solutions are valid in the original equation.	
4 Write the solution.	

9780170464093

MATCHED EXAMPLE 14 | Solving natural logarithmic equations 2

Solve $\log_e(x) + 3\log_e(8) = 7$ for x.

Steps	**Working**
1 Use log laws to express the left-hand side of the equation as a single logarithm.	
2 Change the equation from log form to index form.	
3 Solve the equation expressing the answer in exact form.	

11

MATCHED EXAMPLE 15 | Solving quadratic exponential equations

Solve $e^{2x} - 6e^x + 8 = 0$ for x.

Steps	Working
1 The equation can be written as a quadratic by substituting $a = e^x$.	
2 Solve the equation for a.	
3 Rewrite the solution in terms of x.	
4 Solve the equation, expressing the answers in exact form.	
5 Confirm using CAS. Use e^x, not the letter e.	

TI-Nspire   **ClassPad**

Use Newton's method to approximate a zero for the cubic $y = 2x^3 - 5x + 2$. If $x_0 = -2$, find the value of x_1.

SB

p. 429

Steps	Working
1 Differentiate $y = 2x^3 - 5x + 2$.	
2 Calculate $f(x_0)$ and $f'(x_0)$ where $x_0 = -2$.	
3 Use Newton method to find x_1. $$x_{n+1} = x_n - \frac{f(x_n)}{f'(x_n)}$$	

SB

Using CAS 1:
Newton's method
p. 430

MATCHED EXAMPLE 17	Two iterations of Newton's method

Complete two iterations of Newton's method to solve $x^3 - 5 = 0$ where $x_0 = 1$.

Steps	Working
1 Differentiate $y = x^3 - 5$.	
2 Calculate $f(x_0)$ and $f'(x_0)$ where $x_0 = 1$.	
3 Use Newton's method to find x_1. $$x_{n+1} = x_n - \frac{f(x_n)}{f'(x_n)}$$	
4 Use Newton's method to find x_2. $$x_2 = x_1 - \frac{f(x_1)}{f'(x_1)} \text{ where } x_1 = \frac{7}{3}$$	
5 Calculate $f\left(\frac{7}{3}\right)$ and $f'\left(\frac{7}{3}\right)$.	
6 Substitute in the x_2 formula.	

APPLICATIONS OF DIFFERENTIATION

MATCHED EXAMPLE 1	Differentiating for kinematics

The displacement of a particle travelling in a straight line is given by $x(t) = 3t^3 - 2t$, x in metres and t in seconds.

a Find an expression for $v(t)$.

b Find the acceleration of the particle at $t = 1$.

SB

p. 442

Steps	Working
a Write the function for displacement, then use $\dfrac{dx}{dt}$ to find the velocity.	
b **1** Use $\dfrac{dv}{dt}$ to find a.	
2 Find acceleration at $t = 1$.	

MATCHED EXAMPLE 2 | Positive and negative gradients

State the intervals over which the graph with the rule $y = -x^2 + 1$ has a positive and negative gradient.

Steps	Working
1 Sketch the graph of $y = -x^2 + 1$.	
2 Find the x value of the turning point from the graph.	
3 Identify ↗ or ↘ gradients. Positive for $x < 0$, negative for $x > 0$.	

SB

p. 447

MATCHED EXAMPLE 3 | Strictly increasing and decreasing

State the intervals over which the function with the rule $f(x) = 3x^2 - 2$ for $x \geq -1.5$ is strictly increasing and strictly decreasing.

Steps	Working
1 Sketch the graph of $f(x) = 3x^2 - 2$ with the stated domain.	
2 Find the x value of the turning point from the graph.	
3 From the graph, $f(x)$ is strictly increasing for $x \geq 0$ and strictly decreasing for $-1.5 \leq x \leq 0$.	

MATCHED EXAMPLE 4 Turning points

Find the coordinates of the turning points in the graph $f(x) = \frac{1}{3}x^3 - \frac{1}{2}x^2 - 2x$.

Steps	Working
1 Sketch the graph of $f(x) = \frac{1}{3}x^3 - \frac{1}{2}x^2 - 2x$.	
2 Find $f'(x)$ and solve for zero.	
3 Identify if these are turning points by checking whether the sign of the gradient changes on either side.	
4 Substitute x values into $f(x)$ for coordinates.	

SB

Using CAS 1:
Turning points
p. 451

Find the coordinates of any stationary points of inflection in the graph of $f(x) = -x^4 + 6x^2 - 8x + 3$.

SB
p. 452

Steps	Working
1 Sketch the graph of $f(x) = -x^4 + 6x^2 - 8x + 3$.	
2 Find $f'(x)$ and solve for zero.	
3 Check whether the sign of the gradient stays the same on either side to identify the stationary point of inflection.	
4 Alternatively, find $f''(x)$ and solve for zero.	
5 Find the coordinates of the stationary point of inflection.	

SB

p. 457

MATCHED EXAMPLE 6 | Maximum and minimum points

Find the local maximum and minimum points for the function $f(x) = -(x+1)^2(x-3)^2 + 6$ for the interval $x \in [-2, 4]$.

Steps	Working
1 Sketch the graph of $f(x) = -(x+1)^2(x-3)^2 + 6$ for $x \in [-2, 4]$.	
2 Expand $f(x)$ to find $f'(x)$ and solve for zero.	
3 Identify the local minimum points.	
4 Identify the local maximum points.	
5 Identify the turning points.	

SB

Using CAS 2:
Maximum and
minimum points
p. 458

MATCHED EXAMPLE 7 | Curve sketching

a Sketch the graph of $f: [-4, 5] \to R, f(x) = (x - 2)^2(x + 4)$, labelling key features.

b State the range of f.

Steps	Working
a 1 Explore the general features of the graph.	
2 Find the y-intercept.	
3 Find the x-intercepts.	
4 Find $f'(x)$ and solve for zero for stationary points.	
5 Find the endpoint values of the domain $[-4, 5]$.	
6 Sketch the graph of $f: [-4, 5] \to R, f(x) = (x - 2)^2(x + 4)$.	
b State the range.	

MATCHED EXAMPLE 8 | Maximum value

Find the maximum value of the graph of $f: [-5, 5] \rightarrow R, f(x) = -(x + 2)^2(x - 4)$.

Steps	Working
1 Find endpoint values.	
2 Find the y-intercept.	
3 Find $f'(x)$ and solve for zero for $[-5, 5]$.	

> Note from its equation that this is a cubic function with a leading coefficient of -1 and x-intercepts at 4 and -2.

4 Compare endpoint maximum with local maximum.

5 Use CAS to sketch the graph to check.

6 State the maximum value (at $-5, 81$).

Find the coordinates of the local maximum and minimum points in the graph of $f: [-3, 7] \to R$, $f(x) = \frac{1}{3}(x + 2)^2(x - 6)$.

SB

p. 468

Steps	Working

1 Find $f'(x)$ and solve for zero.

2 Consider the graph shape to decide which is maximum or minimum.

> Note from its equation that this is a cubic function with a leading coefficient of $\frac{1}{3}$ and x-intercepts at -2 and 6.

3 Find the y values of the turning points.

SB

Using CAS 3:
Maximum and
minimum points
p. 469

p. 470

MATCHED EXAMPLE 10 | Maximum and minimum problems

A rectangular cardboard, measuring 7 cm by 4 cm, has square corners of length x units cut out at each corner. These corners are folded up to make a small pencil box. Find the maximum possible volume, in cm^3 correct to one decimal place, for these pencil boxes.

Steps	Working
1 Sketch a diagram.	
2 Set up an equation to describe the volume of the box.	
3 Identify the domain for this problem.	
4 Find $V'(x)$ and solve for zero.	
5 Test whether this is a maximum point by using CAS to sketch a graph or by checking whether the sign of $V'(x)$ changes from positive to negative.	
6 Substitute this value of x into $V(x)$ to find the maximum volume of the pencil box.	

MATCHED EXAMPLE 11 | The anti-derivative

Find an anti-derivative of the function $f(x) = 6x^2 - 2x - 9$.

Steps	Working
1 Write the function as a derivative.	
2 Integrate each term using the formula $\int ax^n \, dx = \frac{ax^{n+1}}{n+1} + c$ ntegratey.	

Using CAS 4:
The anti-derivative
p. 477

Using CAS 5:
The anti-derivative
with a value
p. 478

MATCHED EXAMPLE 12 | Applying the anti-derivative

Find y if $\dfrac{dy}{dx} = x^2 - 2x + 5$ and $x = 3$ when $y = 1$.

Steps	Working
1 Write the expression as a derivative.	
2 Integrate each term and simplify.	
3 Find the value of c by substituting $x = 3$ when $y = 1$.	
4 State the answer including the value of c.	

MATCHED EXAMPLE 13   Anti-differentiating for kinematics

SB

p. 481

The acceleration of a particle travelling in a straight line is given by $a(t) = 3t + 1$.

a Find an expression for the velocity $v(t)$ if $t = 2$ when $v = 0$.

b Find an expression for the displacement $x(t)$ if the particle started at the origin.

Steps	Working
1 State the rule for acceleration.	
2 Use velocity $= v = \int a(t)\,dt$.	
3 Find c using $t = 2$ when $v = 0$.	
4 State the velocity function.	
5 Use $x = \int v(t)\,dt$.	
6 Find d using $t = 0$ when $x = 0$ (starting at the origin).	
7 State the displacement function.	

12

Answers

CHAPTER 1

MATCHED EXAMPLE 1

ba^{-2}

MATCHED EXAMPLE 2

$3^2 \times 7$

MATCHED EXAMPLE 3

$x = 2$

MATCHED EXAMPLE 4

$3x^2 + 10x - 8$

MATCHED EXAMPLE 5

$4x^2 + 4x + 1$

MATCHED EXAMPLE 6

$9x^2 - 36$

MATCHED EXAMPLE 7

$(3x - 2)^2$

MATCHED EXAMPLE 8

a $16x^2 - 9$

b $(8x + 3)(8x - 3)$

MATCHED EXAMPLE 9

a $(x - 2)(x + 3)(x - 3)$

b $(x + 3)(3x - 1)$

MATCHED EXAMPLE 10

$x = -\dfrac{7}{3}$

MATCHED EXAMPLE 11

$(0, 0)$ and $(5, 0)$

MATCHED EXAMPLE 12

$x = 2$ or $x = 3$

MATCHED EXAMPLE 13

$x = -2$ or $x = \dfrac{1}{3}$

MATCHED EXAMPLE 14

$x = -5 \pm \sqrt{\dfrac{31}{2}}$

MATCHED EXAMPLE 15

$x = \dfrac{2 \pm \sqrt{19}}{3}$

MATCHED EXAMPLE 16

a Since $\Delta = 6^2 - 4 \times 3 \times 4 = -12 < 0$, $3x^2 + 6x + 4 = 0$ has no real roots.

b $k \le \dfrac{4}{3}$

MATCHED EXAMPLE 17

$h = \dfrac{A}{2\pi r} - r$

MATCHED EXAMPLE 18

The point of intersection is $(3, 1)$.

MATCHED EXAMPLE 19

The point of intersection is $(5, 3)$.

MATCHED EXAMPLE 20

$k = \dfrac{4}{3}$

MATCHED EXAMPLE 21

$x \ge \dfrac{13}{5}$

CHAPTER 2

MATCHED EXAMPLE 1

$M = \{(2, 3), (2, 1), (4, 3), (4, 5), (6, 6), (6, 1), (8, 5)\}$

Domain $= \{2, 4, 6, 8\}$

Range $= \{1, 3, 5, 6\}$

MATCHED EXAMPLE 2

a Not a function

b Function

c Not a function

d Not a function

MATCHED EXAMPLE 3

a Not a function

b Function

c Not a function

d Function

MATCHED EXAMPLE 4

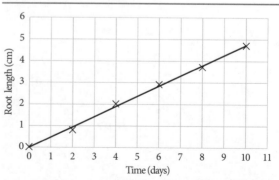

The initial growth rate is about 0.5 cm per day.

9780170464093

MATCHED EXAMPLE 5

a $P(b) = 4n - 220$

b A profit is made when more than 55 are sold.

c

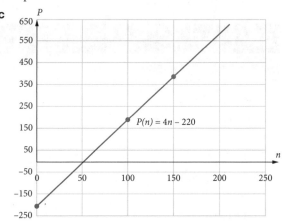

d If all the slices are sold, there is a profit of $580.

MATCHED EXAMPLE 6

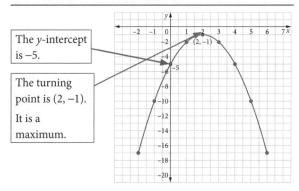

The y-intercept is −5.

The turning point is (2, −1). It is a maximum.

MATCHED EXAMPLE 7

The y-intercept is about (0, 1.2)

The graph does not have any zeros.

There is a minimum about (0, 1.2) and no maximum.

There may be a horizontal asymptote about $y = -1$ and vertical asymptotes at $x = -1.3$ and $x = 0.9$.

The graph does not have any points of inflection.

MATCHED EXAMPLE 8

a The function is 1 : 1, so the inverse is a function.

b The function is not 1 : 1, and the inverse is not a function. Make the domain $x \leq 0$ or $x \geq 0$, so it is 1 : 1; the inverse is a function.

c The function is 1 : 1, so the inverse is a function.

MATCHED EXAMPLE 9

a f^{-1}: (1, 2), (3, −2), (2, 3), (4, −3), (0, 1)

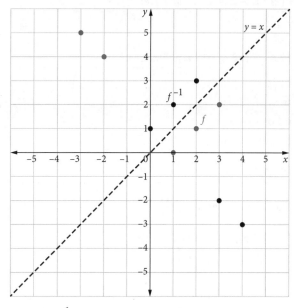

b $f^{-1}(x) = \dfrac{1}{3}x - 4$

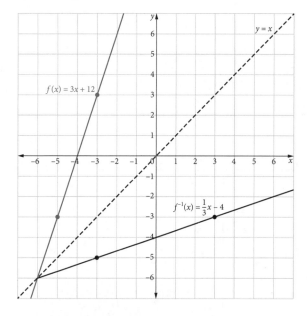

MATCHED EXAMPLE 10

a The maximal domain of $y = \sqrt{x^2 - 1}$ is $[1, \infty)$.

b The maximal domain of $h(x)$ is $R\backslash\{3\}$.

MATCHED EXAMPLE 11

The range of g is $[-9, 15)$.

MATCHED EXAMPLE 12

The range of f is $[4, \infty)$.

MATCHED EXAMPLE 13

The range is $[0, \infty)$ or R^+.

MATCHED EXAMPLE 14

$D = [0, 17)$

MATCHED EXAMPLE 15

a $3x - y + 1 = 0$

b $x + y - 3 = 0$

c $4x - y - 5 = 0$

d $4x + 3y - 16 = 0$

MATCHED EXAMPLE 17

The turning point is $\left(\dfrac{3}{2}, -\dfrac{17}{2} \right)$.

MATCHED EXAMPLE 18

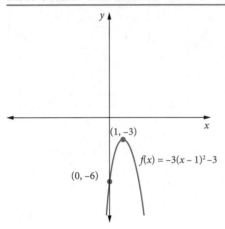

MATCHED EXAMPLE 19

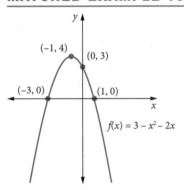

MATCHED EXAMPLE 20

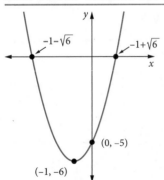

MATCHED EXAMPLE 21

$f(x) = x^2 + bx - 2$ and $y = 5x - b$ have two intersections for $b \in (-\infty, 3) \cup (11, \infty)$.

CHAPTER 3

MATCHED EXAMPLE 1

a $x^2 + x + x^{\frac{1}{2}} - 10$

This is not a polynomial as the power of $x^{\frac{1}{2}}$ is not a whole number.

b $2x^3 + 3x^2 - 8$

This is a polynomial.

Degree = 3

Leading term = $2x^3$

Coefficients = 2, 3, −8

MATCHED EXAMPLE 2

$3x^4 + 3x^3 + 2x^2 - 5x + 2$

This is a polynomial function of degree 4.

MATCHED EXAMPLE 3

a **i** $P(-1) = -4$

 ii $P(a + 2) = 2a^2 + 18a + 32$

b $x = -4$ or $x = -1$

MATCHED EXAMPLE 4

$m = -\dfrac{9}{2}$

MATCHED EXAMPLE 5

a The height of the ball after 5 seconds is 280 m.

b 16.67 s

MATCHED EXAMPLE 6

$a = 4, b = 3$

MATCHED EXAMPLE 7

$(2x^3 - 3x^2 - x + 6) = (x - 1)(2x^2 - x - 2) + 4$

MATCHED EXAMPLE 8

$Q(x) = x - \dfrac{4}{3}$, remainder $= \dfrac{5}{3}$.

MATCHED EXAMPLE 9

$(4x - 1)\left(\dfrac{1}{4}x^2 - \dfrac{7}{16}x + \dfrac{73}{64} \right) + \dfrac{9}{64}$

MATCHED EXAMPLE 10

$3 + \dfrac{16}{x - 4}$

MATCHED EXAMPLE 11

The remainder is –2.

MATCHED EXAMPLE 12

$m = 24$

MATCHED EXAMPLE 13

a $P(-1) = a - b - 5$

   $a - b - 5 = -9$

   $a - b = -4$

b $f(x) = x^3 - 3x^2 + x - 4$

MATCHED EXAMPLE 14

a $P(2) = (2)^3 + 5(2)^2 - 2(2) - 24$

   $= 8 + 20 - 4 - 24$

   $= 0$

   $\therefore x - 2$ is a factor of $P(x)$.

b $2x^3 - x^2 - 7x + 6 = (x - 2)(x^2 + 7x + 12)$

c $2x^3 - x^2 - 7x + 6 = (x - 2)(x + 3)(x + 4)$

MATCHED EXAMPLE 15

$x^3 - 3x^2 - 13x + 15 = (x - 1)(x - 5)(x + 3)$

MATCHED EXAMPLE 16

$P(x) = (x + 1)(x - 2)(x + 3)(x + 2)$

MATCHED EXAMPLE 17

$(x + 1)(2x + 1)(x - 2)$

MATCHED EXAMPLE 18

a $a(3x - y)(9x^2 + 3xy + y^2)$

b $3(x + 3)(x^2 + 3)$

MATCHED EXAMPLE 19

$\pm 1, \pm \dfrac{1}{2}, \pm 2, \pm 4, \pm 8$

MATCHED EXAMPLE 20

$f(x) = \dfrac{1}{2}(2x + 1)\left(x - \dfrac{9 + \sqrt{53}}{2}\right)\left(x - \dfrac{9 - \sqrt{53}}{2}\right)$

MATCHED EXAMPLE 21

$x = 8$

MATCHED EXAMPLE 22

$x = 0$ or $x = 1$ or $x = \dfrac{1}{3}$ or $x = 2$

MATCHED EXAMPLE 23

$x \approx 1.31$ is a root of the equation $2x^3 + 5x - 11 = 0$ in the interval $[1, 2]$.

MATCHED EXAMPLE 24

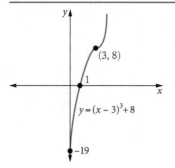

MATCHED EXAMPLE 25

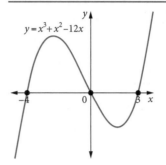

MATCHED EXAMPLE 26

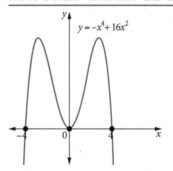

MATCHED EXAMPLE 27

$y = -3(x + 1)^4 - 4$

MATCHED EXAMPLE 1

a

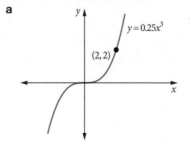

b

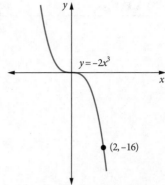

$y = -2x^3$

$(2, -16)$

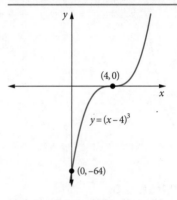

$(4, 0)$

$y = (x - 4)^3$

$(0, -64)$

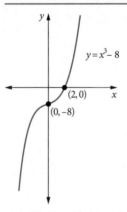

$y = x^3 - 8$

$(2, 0)$

$(0, -8)$

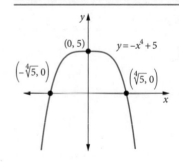

$(0, 5)$

$y = -x^4 + 5$

$\left(-\sqrt[4]{5}, 0\right)$

$\left(\sqrt[4]{5}, 0\right)$

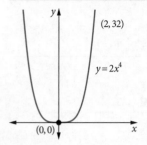

$(2, 32)$

$y = 2x^4$

$(0, 0)$

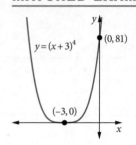

$y = (x + 3)^4$

$(0, 81)$

$(-3, 0)$

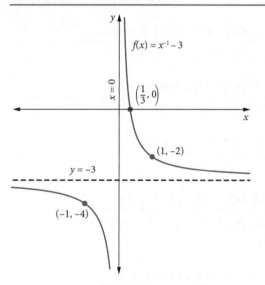

$f(x) = x^{-1} - 3$

$x = 0$

$\left(\dfrac{1}{3}, 0\right)$

$(1, -2)$

$y = -3$

$(-1, -4)$

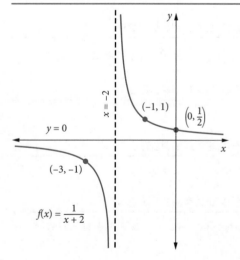

$x = -2$

$(-1, 1)$

$\left(0, \dfrac{1}{2}\right)$

$y = 0$

$(-3, -1)$

$f(x) = \dfrac{1}{x + 2}$

MATCHED EXAMPLE 9

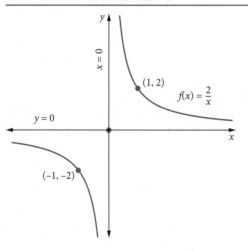

MATCHED EXAMPLE 10

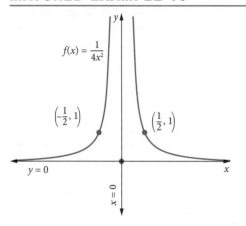

MATCHED EXAMPLE 11

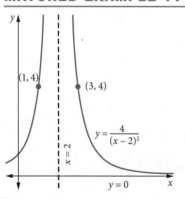

MATCHED EXAMPLE 12

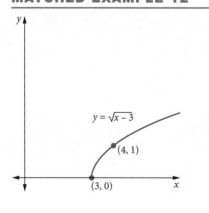

MATCHED EXAMPLE 13

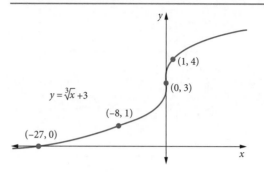

MATCHED EXAMPLE 14

a $g(x) = \dfrac{1}{4}x^2$

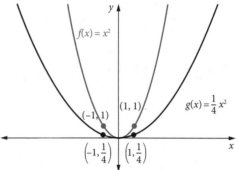

b $g(x) = \sqrt[3]{\dfrac{x}{2}}$

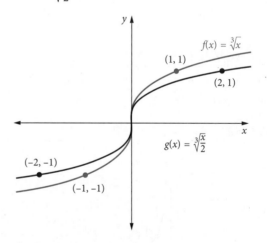

MATCHED EXAMPLE 15

a $g(x) = -\sqrt[3]{x}$

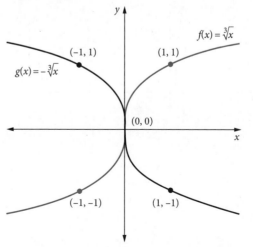

b $g(x) = -\dfrac{1}{x} - 4$

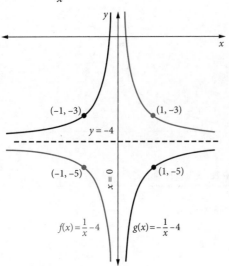

MATCHED EXAMPLE 16

$g(x) = (x+4)^2 - 4$

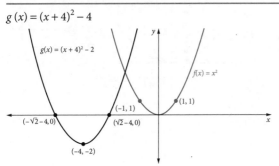

MATCHED EXAMPLE 17

a The new points are $M(-19, -5)$, $N(25, -17)$ and $P(-7, 3)$.

b The new points are $M(2, 2)$, $N(-9, -10)$ and $P(-1, 10)$.

MATCHED EXAMPLE 18

a

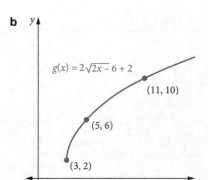

b

MATCHED EXAMPLE 19

a $g(x) = 4[(-x+3)^2 - 2]$

b $g(x) = -\sqrt{2x-1} - 5$

MATCHED EXAMPLE 20

a $f(x)$ has been translated 2 to the left, reflected in the y-axis, translated 3 up and dilated from the x-axis by the factor 2 to obtain $g(x)$.

b $f(x)$ has been translated 8 in the positive x-direction, dilated parallel to the y-axis by the factor $\dfrac{1}{4}$ and translated 2 in the negative y-direction to obtain $g(x)$.

MATCHED EXAMPLE 21

The graph has a dilation from the x-axis by a factor of 3 and translation 1 unit right and 3 units down. Its equation is $y = 3(x-1)^2 - 3$.

MATCHED EXAMPLE 22

g and h are $1:1$ and so have inverse functions, but f is not $1:1$ and does not have an inverse function.

MATCHED EXAMPLE 23

a The inverse is $y = \dfrac{1}{2(x+1)} - \dfrac{3}{2}$.

b $y = \dfrac{1}{2(x+1)} - \dfrac{3}{2}$ is unique, so it is a function. Thus, it can be written as $f^{-1}(x) = \dfrac{1}{2(x+1)} - \dfrac{3}{2}$.

MATCHED EXAMPLE 24

a $y = \dfrac{1}{x+2} - 3$

b

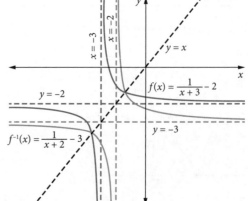

c The graph of $y = \dfrac{1}{x+2} - 3$ is the reflection of the graph of $f(x)$ in the line $y = x$.

d Domain of f is $R\backslash\{-3\}$, and range is $R\backslash\{-2\}$. The inverse relation has domain = $R\backslash\{-2\}$ and range $R\backslash\{-3\}$.

e $f^{-1}: R\backslash\{-2\} \to R, f^{-1}(x) = \dfrac{1}{x+2} - 3$

MATCHED EXAMPLE 25

a

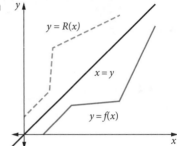

b The inverse is a function since all the y values are unique.

MATCHED EXAMPLE 26

a The dimensions are 3×5.

b $m_{13} = 0$ and $m_{24} = 3$.

MATCHED EXAMPLE 27

1 $\begin{bmatrix} 4 & -8 \\ 12 & 0 \\ 20 & -4 \end{bmatrix}$
3 $\begin{bmatrix} -1 & 2 \\ -3 & 0 \\ -5 & 1 \end{bmatrix}$

2 $\begin{bmatrix} 4 & -2 \\ 7 & 1 \\ 3 & 1 \end{bmatrix}$
4 $\begin{bmatrix} -2 & -2 \\ -1 & -1 \\ 7 & -3 \end{bmatrix}$

5 $\begin{bmatrix} 9 & -6 \\ 17 & 2 \\ 11 & 1 \end{bmatrix}$

MATCHED EXAMPLE 28

$PQ = \begin{bmatrix} 1 & 3 & 5 \\ -2 & -4 & -6 \\ 3 & 4 & 5 \end{bmatrix}, QP = \begin{bmatrix} 6 & 5 \\ -4 & -4 \end{bmatrix}$

MATCHED EXAMPLE 29

The image of $(0, 1)$ is $(3, -1)$.

The image is $y = 2(x-3)^3 - 1$.

MATCHED EXAMPLE 30

The image is $y = \dfrac{\sqrt{x}}{8}$.

MATCHED EXAMPLE 31

The curve is mapped to $y = -x^3 - 4$.

MATCHED EXAMPLE 32

a $T: R^2 \to R^2, T\left(\begin{bmatrix} x \\ y \end{bmatrix}\right) = \begin{bmatrix} 4(3-x) \\ 2+y \end{bmatrix}$

b The curve is mapped to $y = \dfrac{1}{16}(12-x)^2 + 3$.

CHAPTER 5

MATCHED EXAMPLE 1

a $\dfrac{17}{30}$　　　　**b** $\dfrac{1}{2}$

MATCHED EXAMPLE 2

a $\dfrac{1}{6}$　　　　**b** $\dfrac{7}{12}$

MATCHED EXAMPLE 3

a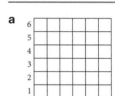

b i $\dfrac{1}{6}$　　　　**ii** $\dfrac{7}{12}$

MATCHED EXAMPLE 4

0.4

MATCHED EXAMPLE 5

$\dfrac{3}{20}$

MATCHED EXAMPLE 6

10

MATCHED EXAMPLE 7

56

MATCHED EXAMPLE 8

a 16　　　　**b** 12

c 28

MATCHED EXAMPLE 9

$\dfrac{1}{12}$

MATCHED EXAMPLE 10

3024

MATCHED EXAMPLE 11

720

MATCHED EXAMPLE 12

a 336

b 360

MATCHED EXAMPLE 13

120

MATCHED EXAMPLE 14

240

MATCHED EXAMPLE 15

$\dfrac{1}{15}$

MATCHED EXAMPLE 16

10

MATCHED EXAMPLE 17

28

MATCHED EXAMPLE 18

56

MATCHED EXAMPLE 19

2940

MATCHED EXAMPLE 20

$\dfrac{1008}{4199}$

MATCHED EXAMPLE 21

a $\dfrac{136}{143}$ **b** $\dfrac{7}{26}$

CHAPTER 6

MATCHED EXAMPLE 1

a The rate of change is 2.5 dollars/kg.

b The rate of change is −5.5 millions/year.

c Since the gradient is not constant, there is no constant rate of change.

MATCHED EXAMPLE 2

a
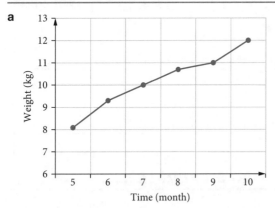

b The points approximately form a straight line with a positive gradient.

c The average growth rate of the puppy is 0.78 kg/month.

d The average growth rate of the puppy is 0.7 kg/month.

This is slightly less than using the first and last data points because the line joining the middle two points is a little 'flatter' than the line joining the endpoints.

MATCHED EXAMPLE 3

The average rate of change between $t = 2$ and $t = 4$ is 8 metres/minute.

MATCHED EXAMPLE 4

The rate of change of radius of the balloon is 0.1125 cm/min.

MATCHED EXAMPLE 5

a 4.1, 4.01, 4.001.   **b** 4

MATCHED EXAMPLE 6

15.84

MATCHED EXAMPLE 7

The approximate rate of change is 17.6.

MATCHED EXAMPLE 8

a 34.481, 32.240 81, 32.024 008

b $\displaystyle\lim_{h\to 0}\dfrac{f(2+h)-f(x)}{h} \to 32$, so the derivative at $x = 2$ is given as $f'(2) = 32$.

c Use CAS to verify your answer.

MATCHED EXAMPLE 9

a $f'(x) \approx \dfrac{f(x+h)-f(x-h)}{2h} = 15$

b $f'(x) \approx \dfrac{f(x)-f(x-h)}{h} = 13.8$

$f'(x) \approx \dfrac{f(x+h)-f(x)}{h} = 16.2$

c The central difference method gave the exact answer. The approach from the left was less than the exact gradient, and the approach from the right gave a greater gradient.

MATCHED EXAMPLE 10

$R(x) = 6x - 0.4$

MATCHED EXAMPLE 11

a The speed was 25 metres/second.

b The truck was stationary for a total of 5 seconds.

c The speed was 20 metres/second.

d The maximum speed was $\dfrac{50}{3}$ metres/second.

MATCHED EXAMPLE 12

a **i** The acceleration is 10 m/s^2.

   ii The acceleration is 0 m/s^2.

   iii The acceleration is −5 m/s^2.

b The steeper the slope, the greater the acceleration. The rate of change of speed is greatest from $t = 7$ to $t = 11$.

c The acceleration is −72 km/h^2.

CHAPTER 7

MATCHED EXAMPLE 1

a 8

b 4

c $\dfrac{1}{2}$

MATCHED EXAMPLE 2

2^{x+2}

MATCHED EXAMPLE 3

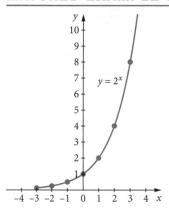

The graph has a y-intercept of 1. The graph's gradient is always positive and is increasing as x increases. The gradient of the graph is also increasing as x increases.

MATCHED EXAMPLE 4

a

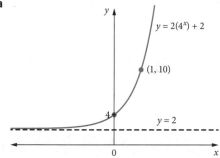

b

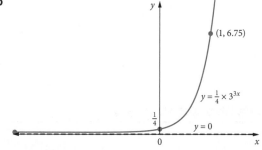

MATCHED EXAMPLE 5

a $c = -2, A = 4$

b $y = 4(3)^x - 2$

MATCHED EXAMPLE 6

a There will be 5 120 000 organisms after 10 days.

b There will be about 905 097 organisms after 180 hours.

MATCHED EXAMPLE 7

After 12 minutes, the amount of water in the tank will be about 155.3 L.

MATCHED EXAMPLE 8

It takes about 90 minutes for the skillet to reach below 21°C.

MATCHED EXAMPLE 9

a $\log_6 (216) = 3$ **b** $\log_4 \dfrac{1}{16} = -2$

MATCHED EXAMPLE 10

a $4^4 = 256$

b $8^{-2} = \dfrac{1}{64}$

MATCHED EXAMPLE 11

a $\log_5 (125) = 3$

b $\log_7 (1) = 0$

c $\log_3 \left(\dfrac{1}{81} \right) = -4$

d $\log_2 (32) = 5$

e $\log_{\frac{1}{6}} (36) = -2$

f $\log_{\frac{1}{4}} \left(\dfrac{1}{16} \right) = 2$

MATCHED EXAMPLE 12

a $\log_2 (0.5) = -1$

b Undefined

c Undefined

d Undefined

e $\log_8 (8) = 1$

f $\log_4 (1) = 0$

MATCHED EXAMPLE 13

a $\dfrac{1}{2}$

b 4

c $\dfrac{2}{3}$

MATCHED EXAMPLE 14

a $\log_2 \left[\dfrac{8x^7}{(y+1)^5} \right]$

b $\log_2 (32x^2)$

MATCHED EXAMPLE 15

$\log_3 (64) \approx 3.786$

MATCHED EXAMPLE 16

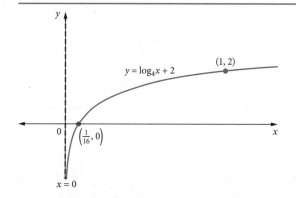

MATCHED EXAMPLE 17

$y = 3 \log_2 (x + 2)$

MATCHED EXAMPLE 18

The magnitude was 5.6.

CHAPTER 8

MATCHED EXAMPLE 1

0.51

MATCHED EXAMPLE 2

$\dfrac{3}{8}$

MATCHED EXAMPLE 3

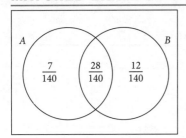

$Pr(A' \cap B) = \dfrac{3}{35}$

MATCHED EXAMPLE 4

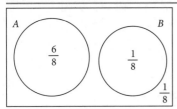

$Pr(A' \cap B') = \dfrac{1}{8}$

MATCHED EXAMPLE 5

$Pr(A \cap B) = 0.43$

MATCHED EXAMPLE 6

$p = 0.11$
$Pr(A' \cap B) = 0.34$

MATCHED EXAMPLE 7

$Pr(A' \cap B') = 0.18$

MATCHED EXAMPLE 8

0.384

MATCHED EXAMPLE 9

0.048

MATCHED EXAMPLE 10

$\dfrac{9}{17}$

MATCHED EXAMPLE 11

0.75

MATCHED EXAMPLE 12

$\dfrac{12}{47}$

MATCHED EXAMPLE 13

0.85

MATCHED EXAMPLE 14

The team wins 22% of its games.

MATCHED EXAMPLE 15

0.42

MATCHED EXAMPLE 16

0.61

MATCHED EXAMPLE 17

$\dfrac{9}{20}$

MATCHED EXAMPLE 18

$\dfrac{7}{13}$

CHAPTER 9

MATCHED EXAMPLE 1

a 270°
b $\dfrac{2\pi}{3}$
c $\dfrac{7\pi}{6}$

MATCHED EXAMPLE 2

a 10.613
b 14.721

MATCHED EXAMPLE 3

a 0.5^c
b 34°

MATCHED EXAMPLE 4

a 3π cm
b 0.45
c $\dfrac{5\pi}{6}$ cm

MATCHED EXAMPLE 5

46°

MATCHED EXAMPLE 6

a $\dfrac{2\sqrt{10}}{7}$

b $-\dfrac{3}{2\sqrt{10}}$

MATCHED EXAMPLE 7

$\sin(\theta) = \dfrac{\sqrt{55}}{8}$

MATCHED EXAMPLE 8

a $\sin(60°)$

b $\cos(20°)$

c $\tan\left(\dfrac{\pi}{10}\right)$

MATCHED EXAMPLE 9

a $-\cos(30°)$

b $-\sin\left(\dfrac{\pi}{4}\right)$

c $-\tan\left(\dfrac{\pi}{3}\right)$

MATCHED EXAMPLE 10

a $-\dfrac{1}{\sqrt{2}}\left(\text{or } -\dfrac{\sqrt{2}}{2}\right)$

b $\dfrac{1}{2}$

c $-\sqrt{3}$

MATCHED EXAMPLE 11

a $y = 3\sin(2x) - 2$ has amplitude 3, period is π and mean value is -2.

b
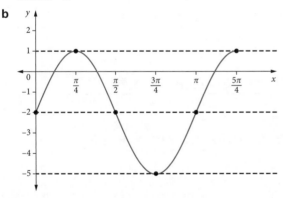

MATCHED EXAMPLE 12

0.67%

MATCHED EXAMPLE 13

a $a = 2$: Dilation from the x-axis (vertically) by a factor of 2.

$n = 3$: Dilation from the y-axis (horizontally) by a factor of $\dfrac{1}{3}\left(\text{period} = \dfrac{2\pi}{3}\right)$.

$c = -2$: Translation 2 units down (vertically, in the negative direction of the y-axis).

b
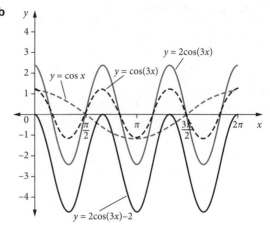

c Domain is $[0, 2\pi]$.

Range is $[-4, 0]$.

MATCHED EXAMPLE 14

a $a = 2$: Dilation from the x-axis (vertically) by a factor of 2.

$n = 2$: Horizontal dilation by a factor of $\dfrac{1}{2}\left(\text{period} = \dfrac{\pi}{2}\right)$.

$c = 1$: Vertical translation 1 unit up.

b Use $n = 2$ to get $x = \dfrac{\pi}{4}, \dfrac{3\pi}{4}$ for $[0, \pi]$.

So $x = \dfrac{\pi}{4}, \dfrac{3\pi}{4}$ are the asymptotes.

c
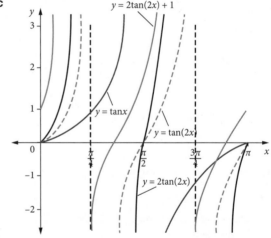

d Domain is $[0, \pi]\backslash\left\{\dfrac{\pi}{4}, \dfrac{3\pi}{4}\right\}$.

Range is R.

MATCHED EXAMPLE 15

a 50 m.

b 6.25 min

c Minimum distance is -50 m, which represents when the point is at ground level, 50 m below the centre of the wheel.

d The t-intercepts represent when the point is level with the centre of the wheel.

MATCHED EXAMPLE 16

a

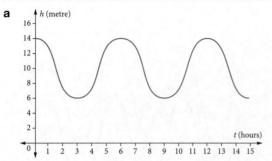

b 12 a.m., 6 a.m. and 12 p.m.

c 6.536 m

CHAPTER 10

MATCHED EXAMPLE 1

−2

MATCHED EXAMPLE 2

There are two different limits, so the limit does not exist as x approaches −1.

MATCHED EXAMPLE 3

a Continuous

b Continuous

MATCHED EXAMPLE 4

$9x^2$

MATCHED EXAMPLE 5

3

MATCHED EXAMPLE 6

a $f(x)$ is differentiable.

b Since $f(x)$ is not continuous at $x = 3$, it is not differentiable at $x = 3$.

MATCHED EXAMPLE 7

a $12x^2 - 3$

b $10x^4 - 0.75x^2 + 12$

MATCHED EXAMPLE 8

a $-\dfrac{4}{v^{\frac{5}{3}}}$

b $\dfrac{2}{3}r + 6r^2$

c $-\dfrac{5}{2}t^{\frac{3}{2}}$

MATCHED EXAMPLE 9

a $\dfrac{65}{32}$ **b** 8

MATCHED EXAMPLE 10

9

MATCHED EXAMPLE 11

108.12

MATCHED EXAMPLE 12

a **i** 57

  ii 97

b The average acceleration is 4 m/s^2.

MATCHED EXAMPLE 13

$y = -3x + 2$

MATCHED EXAMPLE 14

a The gradient of the tangent is 8 at (1, 5) and (−3, 5).

b The tangent is parallel to $y = -x + 3$ at (0, 2) and (−2, 8).

CHAPTER 11

MATCHED EXAMPLE 1

$x = \left\{ \dfrac{7\pi}{6}, \dfrac{11\pi}{6} \right\}$

MATCHED EXAMPLE 2

$x = \left\{ \dfrac{\pi}{6}, \dfrac{\pi}{3}, \dfrac{7\pi}{6}, \dfrac{4\pi}{3} \right\}$

MATCHED EXAMPLE 3

$x = \left\{ -\dfrac{\pi}{8}, -\dfrac{3\pi}{8} \right\}$

MATCHED EXAMPLE 4

$x = \left\{ \dfrac{\pi}{3}, \dfrac{5\pi}{6}, \dfrac{4\pi}{3}, \dfrac{11\pi}{6} \right\}$

MATCHED EXAMPLE 5

$x = 2$

MATCHED EXAMPLE 6

$x = \dfrac{17}{5}$

MATCHED EXAMPLE 7

$x = 2$

MATCHED EXAMPLE 8

$x = 3, 2$

MATCHED EXAMPLE 9

$x = 0, -1$

MATCHED EXAMPLE 10

$x = 10$

MATCHED EXAMPLE 11

$x = \dfrac{105}{16}$

MATCHED EXAMPLE 12

$x = 56$

MATCHED EXAMPLE 13

$x = 2$

MATCHED EXAMPLE 14

$x = \dfrac{e^7}{512}$

MATCHED EXAMPLE 15

$x = \log_e(4), \log_e(2)$

MATCHED EXAMPLE 16

$x_1 = -\dfrac{34}{19}$

MATCHED EXAMPLE 17

$x_1 = \dfrac{7}{3},\ x_2 = 1\dfrac{380}{441}$

CHAPTER 12

MATCHED EXAMPLE 1

a $v(t) = 9t^2 - 2$

b $a(1) = 18$ m/s^2

MATCHED EXAMPLE 2

Gradient is positive for $x \in (-\infty, 0)$.

Gradient is negative for $x \in (0, \infty)$.

MATCHED EXAMPLE 3

Strictly increasing for $x \in [0, \infty)$

Strictly decreasing for $x \in [-1.5, 0]$

MATCHED EXAMPLE 4

$\left(-1, \dfrac{7}{6}\right)$ is a local maximum turning point.

$\left(2, -\dfrac{10}{3}\right)$ is a local minimum turning point.

MATCHED EXAMPLE 5

Stationary point of inflection at $(1, 0)$

MATCHED EXAMPLE 6

The local minimum point is $(1, -10)$.

The local maximum points are $(-1, 6)$ and $(3, 6)$.

MATCHED EXAMPLE 7

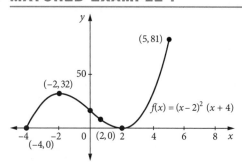

$[0, 81]$

MATCHED EXAMPLE 8

Maximum value $= 81$

MATCHED EXAMPLE 9

Local max $(-2, 0)$ and local min $\left(\dfrac{10}{3}, -\dfrac{2048}{81}\right)$

MATCHED EXAMPLE 10

Maximum possible volume is about 10.4 cm^3.

MATCHED EXAMPLE 11

$y = 2x^3 - x^2 - 9x$

MATCHED EXAMPLE 12

$y = \dfrac{x^3}{3} - x^2 + 5x - 14$

MATCHED EXAMPLE 13

a $v = \dfrac{3}{2}t^2 + t - 8$

b $x = \dfrac{t^3}{2} + \dfrac{t^2}{2} - 8t$